Architect Or Bee?

Architect Or Bee?

THE HUMAN/TECHNOLOGY RELATIONSHIP

by **Mike Cooley**

Introduction by David Noble

Compiled and edited by
Shirley Cooley

SOUTH END PRESS **BOSTON, MA**

© 1980 by Hand and Brain Publications.

All rights reserved. No part of this publication may be reproduced, stored in a retrieval system or transmitted in any form or by any means, without the prior permission of Hand and Brain Publications, 95, Sussex Place, Slough, SL1 1NN, England.

First published in the U.S.A. in 1982 by South End Press.

ISBN 0-89608-131-1
Cover design by Lydia Sargent
Second Printing

Typeset, paste up and design by the collective at
South End Press
302 Columbus Ave
Boston, MA 02116

TABLE OF CONTENTS

Editor's Preface	vii
Introduction by David Noble	xi
1. Identifying the Problem	1
2. The Human Machine Interaction	25
3. Political Implications of New Technology	51
4. Drawing Up the Corporate Plan at Lucas Aerospace	81
5. Some Social and Technological Projections	107

Editor's Preface

It is only recently that the social problems arising from technological change have become the subject of widespread public concern. The predictions that Mike made in his lectures in 1960-70 and thereafter, and in his 1972 book *Computer Aided Design—Its Nature and Implications* are now becoming a reality, and the trade union movement is beginning to recognize that the nature of the new technology is an important and fundamental issue requiring urgent action. The question some are now asking is whether the horse has already bolted and the technology advanced so far along the wrong path that it is too late to divert it along a different one.

This however is not the author's view, or that of the Lucas Aerospace Combine Shop Stewards' Committee which is currently working with other combines, researchers in the field and the Centre for Alternatives Industrial and Technological Systems (CAITS) to produce corporate plans around socially useful products and human centered technologies. The technologies being evolved are those which will again link hand

and brain in the productive process and which will enhance the creativity of the producer.

The book is basically a collection of material extracted from the very wide range of articles, papers, lectures and addresses at trade union meetings which Mike has produced over the last eight years. His work is based on over 20 years experience as an industrial designer in factories where he has felt the impact of new technology on his own job and that of his fellow workers. His analysis has progressed from the use/abuse model evident in his earlier work, to encompass the idea that a society's technology is an integral part of its politics, and that present technological systems are used for purposes of political control by the multinational companies. He uses the computer as a sort of political aperture through which to view the problems advanced technology as a whole is bringing with it. If a society is to be a truly democratic one Mike says, it will have to develop radically different technological systems.

The reader will note that there is an inevitable imbalance in the style and presentation of the material which requires perhaps some explanation. Some parts are closely argued and fully referenced while others are sharp, somewhat rhetorical and not referenced at all. The former material is taken from lectures and papers presented to academic bodies and scientific societies and were produced in the written form in the first place. The sparsely referenced material comes from talks and discussions that were later written up or published in magazines and journals. These include the explanations of the ideas behind the Corporate Plan and the details of some of the products which were the subject of countless meetings within the trade union movement over several years.

In spite of the variation in style, the book stands as a whole in the sense that it is a progression of ideas. It begins by giving the reader an outline of the nature of technological development in industry. It then goes on to demonstrate the effects all this is having on the individuals who are involved with it, those who are displaced by it, and on the society as a whole. It examines the implications of extracting skills from people and programming the fragmented parts into machines and the implications of nuclear power and control by the equipment

Preface ix

including the prospects of a new kind of dual economy already in its infancy.

It is hoped that this book might encourage those who have design, diagnostic and other skills based on the tacit knowledge that Mike talks about, to begin to insist on the development of the kind of systems that the Lucas Aerospace Workers envisage and to organize among themselves to achieve that end. The Lucas Aerospace Combine Shop Stewards' Corporate Plan is based on an entirely new concept. It is a valiant attempt to turn the tide in the direction of socially responsible systems and socially useful products. It sounds so simple, but then most major social innovations do sound simple. It is to the credit of the Lucas workers that this idea is gaining momentum. Their outstanding work is the ground beneath this book. Their courage is ever an inspiration to the author, and it is entirely due to their support and militancy that he finds it possible to talk and write about these matters in public without being victimized.*

Editor

---

*In mid-1981, after a long battle, Lucas Aerospace finally succeeded in terminating Mike Cooley.

Introduction

by David Noble

I first met Mike Cooley the day he came to Boston to give a lecture in a series I was running at M.I.T. on labor's view of technological change. He took the place by storm. It was the first time I had ever seen an M.I.T. audience of smug academics and no-nonsense technocrats respond to a talk with such genuine emotion and with a spontaneous standing ovation. What Mike said is in this book.

For me, the most important thing about Cooley's message, the thing that brought the M.I.T. audience like all others to its feet (a few days earlier he elicited a similar response from the workers at the Ford River Rouge Plant) is that he says what we already know and, by saying it so clearly and so forcefully, permits us to think it. The competence with which he demystifies technical claptrap gives us confidence in our own, quite similar but heretofore silent, heresies. His acknowledgement of the vast resources of so-called "ordinary people" (and especially the tacit knowledge, the "thing we know but cannot tell,") awakens us, technicians and workers alike, to our own untapped talents and capacities. His allegiance to the people with whom he works at Lucas and from whom he derives his own inspiration and strength kindles a kindred spirit in us and empowers us with a renewed sense of our own collective power. Like all great teachers, Cooley is engaged, a participant, and he persuades us less by argumentation than by evocation. Drawing upon his own experience for "evidence," he brings our own experience to consciousness and appeals to it for confirmation, for "proof". And as our own experience resonates with his, it is thereby validated and our perceptions too are confirmed. By his example and that of his fellows at Lucas Aerospace, then, we are encouraged to see what is already there.

In particular, Cooley urges us all to take another look at this thing called "progress," to strip away the technical jargon, to penetrate the ideological haze that clouds our view, to strive to overcome the neurotic compulsions that drive us all (mad), to brave the epithets of the (electronically) elect, the sneers of the officially wise, and to create the space for us once again to be rational. What kind of progress is this? What kind of progress do we want?

A few years ago my mother lost her job to a computer. A legal secretary, she had worked for the same firm for nearly twenty years before being unceremoniously "scrapped" with two days notice and no pension. The computer created jobs for less skilled workers and eliminated those of the more skilled people, like my mother, for whom "retraining" would have meant unlearning. (She was too old to "retool" anyway.) So there she was, home on a Monday morning for the first time in many years, reflecting upon her all too familiar plight. She complained about having no job, about the way she was fired after all those years, about the new workers who do not know half of what she knows, about having no pension and the fact that she wan't getting any younger. But, for all her anger, she was resigned. Shrugging her shoulders, she repeated to herself, as if she had to convince herself, "Well, I guess that's progress."

This story is being repeated with increasing regularity in countless households around the world, as industrialized nations continue their perilous passage through what has been called the second industrial revolution, the computer revolution. In the wake of such progress, countless lives are being disrupted, damaged, destroyed. And yet, stranger than fiction, there is no uprising, no rebellion, even little rethinking. Instead, there is, with few notable exceptions, only desperate accommodation, grudging acceptance of what appears to be inevitable, inexorable. "You can't stand in the way of progress," goes the conventional wisdom, even if it kills you: to do so is irrational and futile. This is the enervating ideology of technological progress, the habit which we have all internalized and of which Mike Cooley is striving to disabuse us. In essence, it is a religious faith, a belief that by replacing labor by machines (mechanization and automation) and reducing reliance upon human skill and effort more things can be produced more cheaply, and that this greater abundance, whatever the human and social cost required to achieve it, constitutes an advance in social welfare. Those who are undone by progress are thus encouraged to believe (and indeed need to believe) that their own misery, however lamentable, is a necessary sacrifice made in the interest of the larger society whose overall happiness will be enlarged as a consequence. Thus, to object is not only futile, it is selfish, parochial, anti-social. Like all ideology, like all faith, this ideology of progress permits no scrutiny and begs all questions. To challenge it is taboo, it is to run the risk of being dismissed without a hearing as a heretic, a Luddite, a fool. Respectability, and even dignity and self-esteem (however alienated), demand compliance. Everyone, after all, is for progress.

Now, we all already know that we are suspicious about this thing called "progress," if for no other reason than the fact that it is being promoted at our expense so stridently, now more than ever, by our bosses, the technocrats, the multinational corporations, the bankers, the military, and the corporate-owned politicians in every industrialized country. But we also want to be taken seriously and to take ourselves seriously, and this ideology of progress, fostered by the interests it camouflages, defines the bounds of respectable discourse, what is to be taken seriously and what is not. Here then is our dilemma: how do we trust and give voice to what we already know from our experience, how do we challenge this false notion of progress and those who profit from the misery it creates in order to survive, and, at the same time preserve some measure of sanity, self-esteem, social support? This is the dilemma Cooley confronts head-on, with unparalleled resourcefulness and acumen. At the heart of his resolve lies the unshakable conviction that human beings are the be all and end all of progress, and not mere resources, factors of production or means of "progress". Before everything else, Cooley reminds us, people come first, for they are the subject of history, not merely the expendable material through which it wends its inexorable way. To the question, who is in charge here, we or somebody's ideas of progress, he responds at once in defiance and affirmation: we are, progress must be our progress or it is not progress at all. How, then, do we take charge of our destiny?

The task entails some enormous undertakings, all the more difficult because they must be engaged in simultaneously. First, we must make it possible to challenge the ideology of progress itself, and there is no better way to do that than by simply beginning the challenge, relentlessly raising the questions which are begged by this idea and avoided by our habits. Whence comes this drive for progress? Who are the people who push and profit from the ideology on which it's based? What gives these people the right to make the decisions that control our lives? Is it state power, property ownership, professional prerogatives? What are the impulses behind this compulsion, the rational parts within this irrational whole? Is it short-run profit, military command control, corporate power and flexibility vis-a-vis competitors and workers, managerial control within the factories, technical delight in remote control and automaticity? Greed, power, enchantment, gadgetry—is this the stuff which will lead willy-nilly to greater social welfare? Is the will-to-progress, for all its scientific pretensions and trappings, really rational? We must carefully re-examine the claims of the enthusiasts: does greater mechanization and automation necessarily result in greater economy, or less? Is this the best way of increasing output, the best way of producing what we need? If greater economy is ever actually attained, is this really equivalent to greater social welfare? Is this the only possible path of technological development, the most sensible the least wasteful, or are there better ways that have not yet been tried? What are those better ways? What are the economic consequences of this current mode of development? Are our economies healthier as a result? Are we more self-reliant, secure, resourceful, versatile, creative, competitive as a result, or less? What are the social and human costs of such supposed progress in terms of unemployment, loss of control over jobs, greater regimentation and routinization of work, destruction of communities, broken health, deteriorating working conditions, the atrophy of irreplaceable human skills and (tacit) knowledge? Is technological development itself, whatever its form, the only or the best solution to our problems, or an escape from our problems? Progress for what? Progress for whom? What kind of progress might we imagine having if we could take charge and steer the course of social change? What would our progress look like and how does it compare with the one we are now creating, mindlessly, in spite of ourselves?

Second, we must go beyond a critical assessment of the current march of progress in order even to make such an assessment. Asking the questions is not enough; to do so forcefully enough we must have some confidence in our answers, and for this we must have a glimpse, at least, of other possibilities. Thus, we must identify and develop the alternatives, the other ways that have been and are still being foreclosed, denied in the compulsive drive to "rationalize" and automate. And to do this, we must tap hitherto unacknowledged resources, knowledge, skills, dreams—in ourselves as well as in others. As Cooley correctly insists, this is not something we can or should do alone; it must be a collective undertaking, an egalitarian, democratic way of establishing new criteria, new specifications, new directions and possibilities, or else we will merely replicate the old in the new. For what is at issue here is far more than new products and processes or better ways of doing better things. We are raising a more fundamental question: what makes a technical design socially viable? According to what criteria, to whose criteria?

Suppose an engineer came up with a new "socio-technical" system for producing something which was elegant, state-of-the-art, apparently reliable and quite promising and one of the operational requirements of the system was that its human "components"—you, the readers of this book, let us say—had to do exactly what the designer instructed them to do, no questions asked. Now, most of the readers of this book, who have better things to do, would no doubt view this engineer as some kind of a nut and dismiss his or her design as impractical and unviable. Yet, designs like this are drawn and put into practice every day. The Pharoah used them. Henry Ford used them. Every corporation in the world uses them today. Here, in such contexts, these silly designs are rendered viable, not by their elegance or the technical prowess of the designer, but by the power relations in which the design is situated. Here, some people, on whose behalf the engineer designs, have the power to get other people to do exactly what they want them to do. The ridiculous design becomes viable and may even be viewed as brilliant. For this reason engineers go to work for corporations; they get the power to realize their designs, without which they would only appear ludicrous. Because the engineer knows that such power to implement a design will be forthcoming, he or she creates designs with that assumption in mind. The end result of this process reflects that assumption as well as the relations of power upon which it rests.

To take another example, suppose an engineer designed a machine for his best friend. When it was finished, he offered it to his friend as a gift, saying with true professional pride, "this is my finest machine; it is so well designed, it can be run by an idiot." No doubt, his friend, who does not consider herself an idiot, would be taken aback: their friendship would for the moment be in doubt, and the engineer would be obliged to redesign the machine for someone who was not an idiot. He would find this very difficult because it goes against all of his training—designing for idiots is the highest expression of the engineering art—and he would quickly discover that he does not know where to begin. (Of course, he might start by pretending that he was designing it for himself.) However, if that same machine had been presented to a manufacturer with the same claim—an everyday occurrence in industry—the engineer would have encountered no such difficulty. The manufacturer knows that he is able to deploy the machine as it is and that, because of his rights as property owner, he can compel his workers to operate the machine and thus do idiot work.

Clearly, then, technical design, like the use of technologies, reflects power relations and class relations. The nature of the relationship between designers and prospective users is "imprinted," so to speak, in the machinery itself. Thus, it matters a great deal whether or not the designers and users are the same people, whether or not they work together as equals and view each other as such, whether or not some have power over others, whether or not all have a voice and a hand in the design process, whether or not designers and users are friends. In short, creating alternative possibilities in design presupposes creating the right social forms in which such possibilities can take shape. And this is precisely what the Lucas Shop Stewards Combine Committee is all about, this working alliance between Lucas workers. The new products, the new work processes that have evolved from their joint efforts could never have been imagined had not the social setting first been created in which designers and users could be unified. This social invention and the experience it made possible gave rise to new technical possibilities. Further, this same social invention, born in struggle and out of necessity and based upon a sound trade union tradition, has provided the necessary social support, the space in which to take another look at progress and the environment in which people could, once again, be rational.

Third, in addition to critically assessing the current drive to progress and the ideology of progress that carries it, and beyond creating the social forms in which to articulate and develop alternative possibilities, it is perhaps most urgent to create a similar space in which to say NO to the calamity now hailed as progress. This might well be the most awesome task of all, but it is the *sine qua non* upon which all else depends. For, unless we can stall this assault upon our organizations, our livelihoods, our communities, our lives, we will not have the chance to develop our strengths, formulate our strategies, identify new possibilities. Somehow, we must make it possible for people to say NO, to refuse to go along and be sacrificed, without their being assailed and dismissed as crazy, selfish, reactionary enemies of progress. Somehow—and this is what Cooley does so superbly—the space must be created in which people can both stand in the way of this so-called progress in order to survive, and still appear to be rational, responsible, forward-looking. Some have tried in the past and failed, but that they attempted it at all, against overwhelming odds, should strengthen our resolve.

In 1932, for example, Dexter Kimball, former president of the American Society of Mechanical Engineers, G.E. manager, and Dean Engineering at Cornell, gave a speech on the meaning of technological progress. In the midst of the depression and popular resentment against engineers for causing "technological unemployment," Kimball acknowledged the problem and urged a shorter work week and ameliorative social legislation. But, he added in his conclusion, "there is also hope in another direction," a slowing down of "the rate of technological progress." "It is obvious," he observed, "that many enterprises have been built far past the point of maximum efficiency; [and] there is always an economic limit to the degree of automaticity that can be applied to productive processes. We may see an era of more moderate sized plants, less automatic," he opined. "It may well be that modern mass production methods carry in themselves the seed of their own limitations, which will eventually tend to limit some of the undesirable social effects. All human experience leads to that belief and perhaps some of the very extreme experiments in mass and automatic production may also contain some salutary lessons for our future guidance." Kimball was talking sense, but he had overstepped the line of respectable discourse, especially the ASME, and had violated the taboo of challenging the ideology of progress. Thus, when the ASME published the speech, this last paragraph was deleted and in its place was substituted, "Finally, whether industrial progress be slow or rapid, these new methods are here to stay." Meanwhile, ASME president Ralph Flanders was in Youngstown, Ohio, exhorting an audience in that once proud steel town, that "there is only one direction in which we can travel, and that is forward, to greater economic, engineering and financial development to obtain the greatest good for the largest number of people." His speech rings hollow now in Youngstown.

A decade and a half later, another man dared to violate the taboo. Norbert Wiener was recognized as a giant among mathematicians and engineers, the father of cybernetics (which wedded computer and servo-mechanisms and paved the way for the automatic factory). But Wiener was not himself a "gadgeteer," the term he used to describe most of his colleagues. He was not so imbued with technical enthusiasm that he failed to see other things. Thus, in 1947 he published a famous letter in which he announced that he would no longer publish his work out of fear that it would inevitably get into the hands of "irresponsible militarists." Two years later Wiener acted again. He had become deeply concerned about the probable social implications of the applications of computers and servo-mechanisms to industry, expecially the dire consequences for labor, and he steadfastly had refused to do any consulting work for companies getting into this field, such as General Electric. In 1949, in desperation, he wrote to Walter Reuther of the UAW. He told Reuther about his fears of massive unemployment and the reduction of the working population to slave labor conditions, warning him of "the very pressing menace of large-scale replacement of labor by machine. I do not wish to contribute in any way to selling labor down the river," he told Reuther, and "for me merely to remain aloof is to make sure that the development of these ideas" will lead to disaster. "You might," therefore, "want to steal a march upon the existing corporations to insure that the advances in the technology benefit labor," Wiener suggested, "or it may be that you think the complete suppression of these ideas is in order. In either case," he pledged, "I am willing to back you loyally...in what I consider to be a matter of public policy." Reuther took Wiener's advice and launched an early effort to try to deal with the problems of automation. But an expanding economy absorbed some of the unemployment Wiener predicted, the diffusion of the new technology was much slower than had been anticipated, public attention turned elsewhere, and debates over the automation were prematurely closed. The calamity, as it turned out, was delayed—until now. Meanwhile, at the point in his career when Wiener began to share his alarm about the social consequences of so-called progress, people started to talk about his approaching senility.

And finally, most recently, there is John Parsons, the inventor of numerical control and, as such, the originator of computer-based manufacturing. He is the man whom the Society of Manufacturing Engineers has recognized as the father of the second industrial revolution. In 1965 the editor of *American Machinist* interviewed Parsons and asked him, as a major contributor to modern industry, what he thought should be the best technological path to travel over the next ten years. Parsons replied, to his interviewer's astonishment, that he thought there should be a moratorium on all new technological development, that instead of developing new gadgets we should learn first how to use more effectively what we already have. More important, he stressed that, rather than trying to develop technical solutions to our problems, we should get to work trying to solve them in more promising ways, such as by improving labor relations. Here, Parsons opined, is where the real gains are to be made. "A moratorium!" the editor exclaimed, "if I write that in my magazine everyone will think I'm crazy." Parsons suggested instead that the editor simply quote him as saying it. "But then they'll think you're crazy," he replied. No doubt to protect Parson's reputation, as well as his own, the editor never printed the interview.

These three examples illustrates something that is still quite rare in our society and which is now more urgently needed than ever. These three men, none of whom could be called anti-technologists or machine-breakers, had, for a moment at least, seen their way clear of the suffocating ideology of progress. Soberly, responsibly, rationally, they suggested, for rather different reasons, that we stop and take stock, slow down, rethink where we are headed. But the hegemonic ideology of progress prevailed and their important reflections appeared absurd. It is this appearance of rationality as absurd that we must strive to overcome. Together, in our thinking and acting we must create the space to say NO to a progress that is killing us in order to say YES to a progress that will more surely enhance human dignity, creativity and happiness.

Finally, new possibilities will not emerge from contemplation alone nor even from active experimentation, but from ongoing political struggle—in our workplace, our unions, our communities, our government, our streets. We must begin to empower ourselves to be in control of our fate, to be the creators that we are, and to struggle against those who, in the name of their progress, deny us ours. Above all, we must strive to create the social environment in which these new possibilities become not simply viable but necessary, in which the luxury of denying them at our expense becomes politically impossible. It is no accident that the initiative for this movement should come from workers who faced economic disaster when they were threatened with the sudden loss of their livelihoods. For them, the search for new possibilities had already become a necessity, a matter of survival. Now, the Lucas workers through their example are inspiring other workers in other countries who are similarly faced with a threat to their survival to turn their necessity into invention. Through their efforts and ours, such invention might yet alter the course of progress. Mike Cooley's book, like the experience that it embodies, takes a giant step in that direction.

ONE

Identifying the Problem

There is still a widespread belief that automation, computerization and the use of robotic devices will free human beings from soul destroying, routine, backbreaking tasks and leave them free to engage in more creative work. It is further suggested that this is automatically going to lead to a shorter working week, longer holidays and more leisure time—that in an all round way it is going to result in an "improvement in the quality of life." It is usually added, as a sort of occupational bonus, that the masses of data we will have available to us from computers will make our decisions so much more creative, scientific and logical, and that as a result we will have a more rational form of society.[1]

I want to question some of these assumptions and attempt to show that we are beginning to repeat in the field of intellectual work most of the mistakes already made in the field of skilled manual work at an earlier historical stage when it was subjected to the use of high capital equipment. I move from manual to intellectual work quite deliberately because I resent the division between the two, and I therefore draw parallels throughout.

In my view it would be a mistake to regard the computer as an isolated phenomenon. It is necessary to see it as part of a technological continuum discernible over the last 400 years or so. I see it as another means of production, and as such, it has to be viewed in the context of the political, ideological and cultural assumptions of the society that has given rise to it. Consequently, I look critically at technological change as a whole in order to provide the framework for questioning the way computers are used today. I take the Hegelian view that truth lies in the totality, and therefore after considering some of the equipment currently in use, I will relate its effects to the labor process and try to give an overall view of what is happening. The equipment and processes described are not necessarily the most advanced or the latest in their field. They are chosen because they are typical of the kind of changes that are taking place in design.[2] The problems I see arising within the design activity in the areas described will be universally applicable whether one is talking about computers in insurance, banks, newsprint industry or any other field.

THE EQUIPMENT

The first piece of equipment considered is one which is being used to replace the function that was traditionally known as drafting. Up to the 1940s the draftsperson was the center of the design activity. He could design a component, draw it, stress it out, specify the material for it and the lubrication required. Nowadays, each of these is fragmented down to isolated functions. The designer designs, the draftsperson draws, the metallurgist specifies the material, the stress analyst analyzes the structure and the tribologist specifies the lubrication. Each of these fragmented parts can be taken over by equipment such as this automatic drafting equipment. (*Fig. 1*)

With this equipment, the draftsperson no longer needs to produce a drawing and so the subtle interplay of interpretation and modification as the commodity was being designed and related to the skilled manual workers on the shop floor is being ruptured. What the draftsperson now does is work on the digitizer and input the material through a graticule or teletype. An exact reading is set of the length of each line, the tolerance and other details. The design comes out as a tape which is expanded in

Identifying the Problem 3

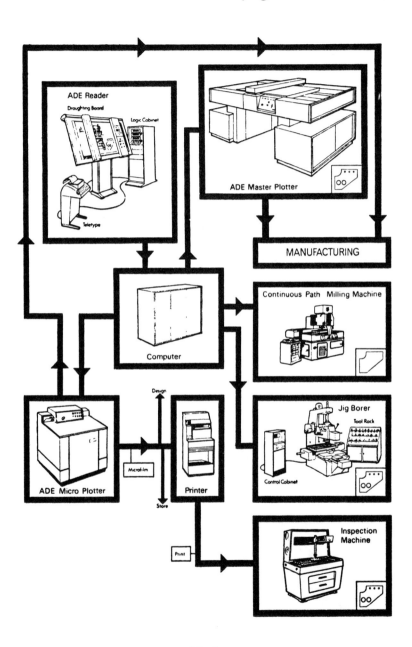

Fig 1.
ADE System for engineering data processing

the computer after which it operates some item of equipment such as a jig borer or a continuous path milling machine. After that, the equipment itself will do the inspecting. If perchance you want a drawing in order to show the customers exactly what they are purchasing—and that's the only reason you would bother to do it—then you can produce one on a master plotter very accurately. You can get a less accurate one on the microplotter which also produces an aperture card.

What is important in all this is not only that the fragmented functions of the designer have been built into the computer, but the highly skilled and satisfying work on the shop floor has also been destroyed. It is no longer a question of supply and demand, of a slump or a boom; these jobs have been technologically eliminated and yet they were some of the most satisfying and fulfilling jobs on the shop floor.

Quite apart from the suffering of those involved and the destruction of the creativity the worker used in doing the job, what must be of concern to all of us is where the next generation of skills is coming from—skills which will need to be embodied in further levels of machines. The feel for the physical world about us is being lost due to the intervention of computerized equipment and work is becoming an abstraction from the real world. In my view, profound problems face us in the coming years due to this process.[3]

Part of the skill of a draftsperson or a designer was the ability to look at a drawing and conceptualize what the product would look like in practice. That conceptualization process is now also being eliminated by computers. I have in my lectures illustrated systems which are capable of tracing round the profile of the conventional type drawing which includes plan and elevation views, and produce an accurate three dimensional representation of the object on the screen before it actually exists in practice. The computer will rotate it through any angle for you when given instructions. This can be extended further in the field of architecture, for example. A visual display such as the one described could be made of any proposed municipal building, and local people could be involved in deciding whether they approved of its design and its location. Normally, a plan of a proposed municipal building is available for inspection in the town hall, but for most people this means very little. It is intelligible only to an elite group.

By sensitive use of the computer in this way we could actually reverse the process and involve the community in deciding the kind of buildings it wanted.[4]

Theoretically then, there is the potential for democratizing the decision making process. I will argue elsewhere, however, that the computer is in fact used to reinforce the power of minorities over majorities. There is a great danger that the whole thing can be extremely manipulative. If you have a perspective view of a building on a visual display unit and you take the point of convergence far away, you can make the building look slim and attractive, disappearing into the horizon. On the other hand, if you take it close up you can make the building look like a high rise block. Thus it is very easy to manipulate public opinion, and I think that some architects are not beyond that sort of thing!

At an even higher level you can get what appears to be all the power of retrospective logic. Anyone who has worked as a designer will know that you get your best ideas afterwards when you can see the kinds of mistakes you have made while designing. I have described systems now used in the field of architecture which aim to provide the designer with some kind of retrospective logic. They were adapted from visual simulation techniques used to train astronauts in docking manoeuvres. The underlying principle is that images are presented at 30 times per second on a color visual display unit. Standard cues of depth are given as overlapping surfaces, and the apparent size of the object is given as inversely proportional to the distance from the observer. This is, of course, the typical Western cultural way of presenting visual data of this kind.

In the case of architectural design, each building and object is defined in its own three dimensional co-ordinate system. These are then presented as a hierarchical structure of co-ordinate data. This means that all the existing buildings can be input as data structure, and the new building to be designed is shown within the context of the existing architectural arrangements. That is to say one can experience in the equivalent of "real time" walking towards a building that still does not exist in practice. One can experience the sensation of going inside the proposed building and looking out at the existing buildings. One can take windows out, move them about, enlarge the whole thing and take it right outside the proposed site. The aim is to assess the total effect of the new building on the whole environment before constructing it. There are already grounds for believing, however, that images of reality as presented in that form are still very different from the actuality. When the building is erected you can get a ghetto-like prison atmosphere which is not apparent when you have severed the relationship between the object and the real world.

In the past, skilled workers have had in the main a tacit understanding of mathematics through their ability to analyze the size and shape of components by actually working on them.[5] More and more, that knowledge has been abstracted away from the labor process and has been rarified into mathematical functions. In diagrammatical form the functions might be of the kind shown in the figure (*Figure 2*). The first one is a Bessel function and the second a sinusoidal function. The latter might represent the way a shaft is vibrating, for example.

Now, if these kinds of mathematical problems were displayed diagrammatically where they could be used by the skilled worker, he or she would now be able to be involved in making design decisions in just the same way as a skilled prototype fitter would have done in the past by looking at the behavior of the actual equipment. At present, that knowledge has been taken away and rarified into a computer program where only a small group of people can work on it. What would have gone on in reality is conceptualized, objectivized and fed to the computer. The reverse should be the case in my view, and I can see quite specific political and other reasons for doing it. In its present mode of usage the computer is a tool for silencing the common sense and creativity of the skilled worker on the shop floor.

A major application of Manned Computer Graphics is in the field of structural analysis. Equations required for the analysis of the structure are automatically set up and are solved automatically upon request of the analytical output. Displacement, loads, shear and moments are computed and conveniently displayed for perusal. Changes of input conditions are easily facilitated, and the corresponding output is displayed upon request. Constraining

Identifying the Problem 7

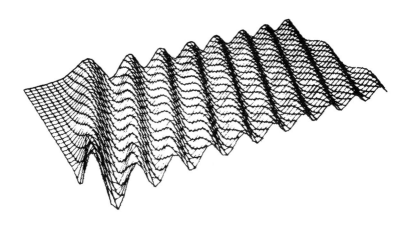

$$J_n^{(x)} = \frac{F_n^{(x)}}{\alpha}$$ Where: n0→10 x0→60

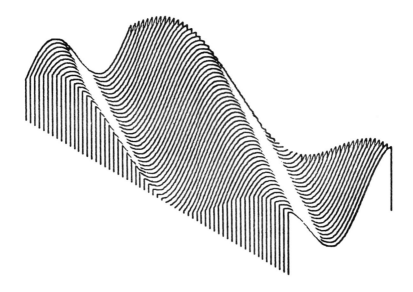

Computer produced solution space surface for SIN $(8*(X-1)/X_L+1/4$
$(Y-1)+1.0$

Fig 2

forces may be placed by using a light pen. *Fig. 3* shows the exaggerated displacement under load.

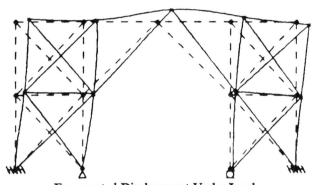

Exaggerated Displacement Under Load

Fig. 3

This equipment represents a deskillling process because it becomes possible to use designers and stress analysts with much less ability and experience than was previously required.

Now the windloading on a tower is a quite complex analysis problem *(Fig. 4)*. The stress in the structure as it distorts can be

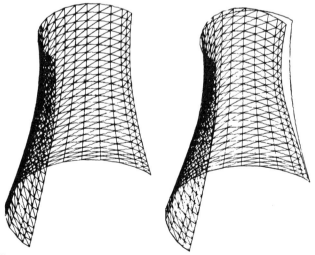

The illustrations are finite element idealisations of a sectioned cooling tower. The right hand picture showing the exaggerated distortion under wind loading.

Fig 4

obtained from a computer package. The distortion is represented on a screen as shown. What all this represents, in fact, is that the knowledge which previously existed in the consciousness of the stress analyst, which was his or her knowledge, taken home every night and which was part of that person's bargaining power, has now been extracted from them. It has been absorbed and objectivized into the machine through the intervention of the computer and is now the property of the employer, so the employer now appropriates part of the worker himself through the intervention of the computer and not just the surplus value of the product. Thus we can say that the worker has conferred "life" on the machine and the more s/he gives to the machine the less there is left of the worker!

Another possibility for the computer, as some architectural readers will know, is to analyze a whole range of variables that exist in an application and plot the solution. This has been used by a planning agency that required a layout of villas on an island. The layout was to be such that each villa had the same amount of sunlight, garden space, view of the sea and many other variables. The computer handled this by doing an initial layout and gradually rearranging and modifying it to fit in with the terrain until it ended up with a final layout superimposed upon the map of the island. This enabled a very dense distribution of buildings, which in my view was a grotesque thing to do to an island, but does show the power of the computer. This too was very early work. There are much more sophisticated packages available today.

In the medical field there are several uses of computer aided design which in my view are positive, although they do bring with them a whole range of problems which I shall describe later.

One example is the use of a Visual Display Unit in the design of equipment for ear protection. The V.D.U. will display the form of soundwaves in the inner ear so your protective equipment can be modified until certain sounds are shown to have been eliminated. Theoretically, you could design ear protection equipment that would allow human speech through and eliminate all other forms of noise. You could in fact choose what you want to hear.

A second example is the use of a V.D.U. in the design of artificial limbs. A graphic system will work out the area of the kneecap joint required for the particular individual for whom it is being designed. The whole structure can be animated on the screen. The person who was to use the limb could be involved in discussing its design before it was actually made.

A third example is the use of computer aided design (C.A.D.) in the design of heart valves. Techniques originally developed to display characteristics in hydraulic circuits in aircraft are used to display on the visual display unit, the venturi and other phenomena as the blood flows through the heart valve. Working interactively, it is possible to modify the valve orifice diameters and other critical physical dimensions and display on the screen the resultant flow characteristics. It is thus possible to optimize the heart valve design to meet the special requirements of the individual patient.

When one considers all these uses for the computerized equipment one gets the immediate impression that it must automatically improve the whole creativity of the designer concerned. However, there are enormous problems involved which require discussion. The complex communications that go on between human beings during problem solving activities are being distanced by the computer and by the systems interfacing the people with the computer with attendant consequences that are very serious and far reaching. Look at the job of a building designer for example. In the past, when designing a building, s/he would go out to the site to see how the structure was progressing. S/he would discuss it with the site engineer and maybe modify the design. Now it is possible to have a display on the site so that visits are unnecessary because the designer and engineer can have an abstracted conversation via the equipment. The designer's drawings will be transmitted through telephone lines and then displayed on the screen so the physical contact between the designer and the site is cut out. Apart from the design implications, the system will tie people down to the machine more and more and the break of getting away from the drawing office and on to the site, which was always one of the perks of the job, will no longer be acceptable.

In spite of the power of this equipment to do some really good work, it brings in its wake all the problems which high capital equipment brought to manual work at an earlier historical stage. First, it shares with all other equipment historically an ever increasing rate of obsolescence. Wheeled transport existed in its primitive form for thousands of years. Watt's steam engine was working for over 100 years after it was built. High capital equipment in the 1930s was written off after 25 years and equipment of this latest kind will be obsolete in three or four years time. Economists would say that this shows the increasing short life of fixed capital.

Second, when viewed historically, it will be seen that the total cost of the means of production is ever increasing. This is in spite of the reduction in the cost of hardware. While these costs are reduced dramatically as computerized systems are miniaturized, the total cost of the system, including the plant and the processes which the hardware is used to control, is ever increasing. The most complicated lathe one could get 100 years ago would have cost the equivalent of ten workers' wages per annum. Today, a lathe of comparable complexity, with its computer control and the total environment necessary for the preparation of the tapes and the operation of the machine, will cost something in the order of a hundred workers' wages per annum. This is frequently forgotten when people talk about microprocessors, and one is almost given the impression that you could fly the Atlantic on a chip, that you could excavate the foundations of buildings or process chemicals (even food) with microprocessors in isolation!

RATE OF CHANGE

A discernible feature about modern equipment of any kind is the rate of change that is now driving us along at an incredible tempo. Over the last century alone the speed of communication has increased by 10^7, of travel by 10^2, of data handling by 10^6. Over the same period energy resources have increased by 10^3 and weapon power by 10^6. We are being drawn along in this tremendous technological inferno, and it means that the knowledge we have and the basis upon which we judge the world about us is becoming obsolete at an ever increasing rate, just like the equipment. It is now the case in many fields of endeavor that simply to stand still, you have to spend 15% of your time updating your knowledge. The problems for older workers are really enormous. There is a mathematical model justifying this.

Suppose S represents the total stock of useful theoretical knowledge possessed by an engineer.

F the fraction of this knowledge which becomes obsolete each year.

R the fraction of his working time devoted to acquiring fresh theoretical knowledge.

L his learning rate.

Then   LR=FS.

Assuming S is constant and equal to the stock of knowledge with which the engineer left university and that his/her average rate of learning remains the same as it was during his/her four year university course, and assuming also that 5% of that knowledge becomes obsolete each year, the equation becomes:

$R \frac{S}{5} = 0.05 S$ where R=0.15 or 15% of working time.

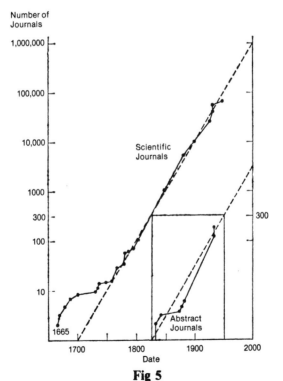

Fig 5

Total number of Scientific Journals and Abstract Journals founded, as a function of date.

Identifying the Problem 13

As might well be expected, the number of journals to be studied is ever increasing. This was shown by Hilary and Steven Rose in *Science and Society (Fig. 5)*.

In some fields, the rate of obsolescence is much greater than that indicated above, particularly in certain areas of computer application. Norman Macrae, Deputy Editor of the *Economist*, stated in the issue of 22 January 1972: "The speed of technological advance has been so tremendous during the past decade that the useful life of the knowledge of many of those trained to use computers has been about three years." He further estimated that "a man who is successful enough to reach a fairly busy job at the age of 30, so busy that he cannot take sabbatical periods for study, is likely, by the age of 60 to have only about one eighth of the scientific (including business scientific) knowledge that he ought to have for proper functioning in his job."

It has been said that if you could divide knowledge into quartiles of outdatedness, all those over the age of 40 would be in the same quartile as Pythagoras and Archimedes. This alone shows the incredible rate of change, and the stress it places upon design staff, particularly the older ones, should not be underestimated. What is happening is that the organic composition of capital is being changed. Processes are becoming capital intensive rather that labor intensive. As a result of technological change we are moving towards a form of society depicted in the following diagram *(Fig. 6)*.

It shows that around 86% of the population of the U.S.A. were involved in agriculture in the early 1800s. This was subjected to mechanization, the use of chemicals and now automation so that now only 6% of the population produce a far greater agricultural output than in 1800. There are automatic tractors that can feel their way around a field so that no human being is required. During the same period, manufacturing industry was growing. Up to the 1950s and certainly the early '60s, it was subjected increasingly to mechanization and automation. The proportion employed in manufacturing is now reduced to about 36% and declining rapidly, while at the same time, the white collar administrative, management and scientific area has been growing.

That in turn is now being subjected to massive computerization and automation which will do to this area what has been

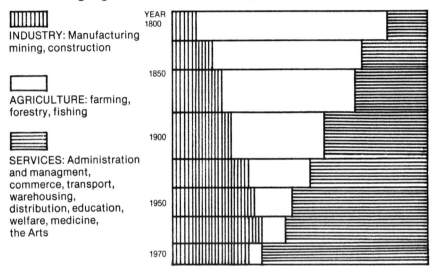

Fig. 6

done to the others. We are confronted therefore with massive and growing structural unemployment. More and more we are moving into a position where large numbers of people are going to be denied the right to work at all.

The non-availability of work for large numbers of people may not seem too great a tragedy to some. I think it is necessary to declare that I believe that work is very important to people. Not the grotesque alienated work which has developed over the last 50 years, but work in its historical sense which links hand and brain and which is creative and fulfilling. If you ask anybody what they are they never say "I am a Beethoven lover" "...a Bob Dylan fan" or "...a James Joyce reader". They always say "I am a fitter" "...a nurse" "...a teacher" and so on. We express ourselves through our work. We relate to society through our work and we are creative through our work.

Clive Jenkins has described the decline of the Protestant ethic. Now I do not support the Protestant ethic, but I do make the point that work in its historical context is important to society and we are increasingly being denied the opportunity to do that work.

There is more to it than that, however, for those who are displaced by technological change are not the only ones seriously affected. What is happening to those remaining in work is really worth analyzing.

THE PRESSURE IS ON!

It is widely recognized on the shop floor that technological change has resulted in a frantic work tempo for those who remain. At the Triumph Plant in Coventry, England, it is reckoned that a human being is "burned up" in ten years when working on the main track. The Engineering Union, to which I belong, was asked to agree that nobody would be recruited over the age of 30 so that the last ten years would be from 30 to 40. The same kind of thing is happening in parts of the steel industry. The workers there have become party to an agreement which includes a medical check.

Now, in a civilized society a medical check would be an excellent thing. If something were the matter with you it would be discovered, put right, and you would continue working. *This medical check is a kind of industrial M.O.T. test.** Your response rate is worked out (like a diode) to see whether or not you are fast enough at interfacing with the equipment. If you fail, you are put on to second or third rate work. There is an established list of wages for those who have been so replaced because their reaction time was not fast enough.

For those who do not work in the automotive industry, it is difficult to appreciate how bad the situation is becoming and to what extent workers are being paced by these computerized, high technology systems. In the section where they press out the car bodies in one car company, workers are subject to an agreement on the makeup of their rest allowance. The elements are as follows:

---

*This is a compulsory annual functional test given by the Ministry of Transport for all automobiles in the U.K. to ensure that they are 'roadworthy'.

Trips to the lavatory	1.62 minutes. It is computer precise; not 1.6 or 1.7 but 1.62!
For fatigue	1.3 minutes
Sitting down after standing too long	65 seconds
For monotony	32 seconds—and so the grotesque litany goes on.[6]

The methods engineers located the toilets strategically close to the production line so that operators could literally flash in and flash out. What arrogance some technologist had to be able to do that to another human being! If we have strikes in the automotive industry we must not be surprised. In my view they are right to strike against conditions of this kind, yet all the time this is the kind of philosophy behind the design of much of the equipment produced for industry today.

SEETHING INDUSTRIAL DISCONTENT

Pronouncements on the dehumanization of work in so-called technologically advanced societies have tended to concentrate on manual tasks. This is not surprising since despotism in the factory is now so great as to be counter productive. In addition to well known problems in the United States and in Great Britain, there are those experienced in other industrial nations. At Fiat in Italy the absentee rate is 18%. In Sweden the government has introduced protective workshops for those who need protecting from the advanced technology which we had always been given to understand would liberate all of us. Along with these examples, the sabotage of products at the General Motors plant in Lordstown reveals but the tip of a great iceberg of seething industrial discontent.

Less spectacular, but even more significant as indicators, are the ever increasing rate of production defects and errors, the widespread increase in accidents, absenteeism and turnover, and the very real difficulty, in spite of the bait of a financial anesthetic, in finding adequate numbers of workers to submit to the degradation of the modern factory.

Even when the employer does succeed in finding sufficient "human machine appendages", his/her problems are by no means at an end. The industrial worker, despite a class-ridden educational system which systematically seeks to reduce his or her expectations to an absolute minimum, and despite the continual bludgeoning by the mass media, still retains a degree of dignity and initiative which employers find alarming. Indeed, it is one of the greatest tributes to human dignity that the industrial worker obstinately refuses to meet the specification "That he should be so stupid and so phlegmatic that he more nearly resembles in his mental makeup the ox, than any other type" (*Taylor* 1947).

It is not surprising that human beings, when viewed in this way, and when required to work within a productive process which treats them as oxen, should take what steps they can, however defensive, to assert their humanity. These attempts are not unrelated to the failure rate in parts of industry. It has reached such proportions that something like half the equipment lies idle at General Motors' most modern factory "Where the intensity and monotony of work surpasses anything previously imposed on assembly line workers" (*Gorz* 1976).

Faced with this massive and growing contradiction, it is not surprising that employers are seeking aid from a whole host of "Hawthorne Agents" such as job enrichment specialists, group technologists and industrial psychologists. The industrial reality is that these "agents" in no way change the basic power relationships which give rise to these contradictions in the first place. It is, as a Lucas shop steward put it "Like keeping people in a cage and debating with them the color of the bars."

It may be felt that although this can happen in the field of manual work, it cannot occur in the field of intellectual work because you cannot do that to intellectual work.

In my view, the computer is the Trojan Horse with which Taylorism is going to be introduced into intellectual work. When a human being interacts with a machine, the interaction is between two dialectical opposites. The human is slow, inconsistent, unreliable but highly creative, whereas the machine is fast, reliable but totally non-creative.[7]

Originally, it was held that these opposite characteristics— the creative and the non-creative—were complementary and would provide for a perfect human/machine symbiosis, for example, in the field of Computer Aided Design.[8] However, it is not true that design methodology is such that it can be separated into two disconnected elements which can then be combined at some particular point like a chemical compound. The process by which these two dialectical opposites are united by the designer to produce a new whole is a complex, and as yet ill defined and researched area. The sequential basis on which the elements interact is of extreme importance.

The nature of that sequential interaction, and indeed the ratio of the quantitative to the qualitative depends upon the commodity under design consideration. Even where an attempt is made to define the proportion of the work that is creative, and the proportion that is non-creative, what cannot readily be stated is the stage at which the creative element has to be introduced when a certain stage of the non-creative work has been completed. The very subtle process by which the designer reviews the quantitative information assembled and then makes the qualitative judgement is extremely complex. Those who seek to introduce computerized equipment into this interaction attempt to suggest that the quantitative and the qualitative can be arbitrarily divided and that the computer can handle the quantitative.

The speed at which computers are capable of carrying out immense computations is almost impossible to grasp. For example, to calculate all the stresses in the Gyretron—the space frame center piece of the 1967 International Trade Exhibition, EXPO 67—a computer was employed for two hours. A mathematical graduate could have performed the same calculations but would have taken about 30,000 years! This is equivalent to about 1,000 mathematicians working for their entire lifetimes.

THE FASTER THE BETTER!

Where computerized systems like this are installed, the operators are subjected to work which is alienating, fragmented and of an ever increasing tempo. As the human being tries to keep pace with the rate at which the computer can handle the quantitative data in order to be able to make the qualitative value judgements, the resulting stress is enormous. Some systems we have looked at increase the decision making rate by 1800 or 1900%, and work done by Bernholz in Canada has shown that getting a designer to interact in this way will mean that the designer's creativity, or ability to deal with new problems, is reduced by 30% in the first

Identifying the Problem 19

hour, by 80% in the second hour, and thereafter the designer is shattered![9] The crude introduction of computers into the design activity in keeping with the Western ethic "the faster the better" may well result in a plummeting of the quality of design.

There are arrangements in some systems where there is a set length time for handling the data (17 seconds is an example). If you do not comply with this you are downgraded to "head-scratching status" as they call it. The anxiety of those involved can be measured, for they display all the signs of stress such as perspiration, higher pulse rate and increased heartbeat. Suppose the image is about to disappear from the screen and you haven't finished with it. You can hold or recall it, but everyone in the office knows when you have become a headscratcher. You are being paced by the machine, and the pace at which you work is becoming more and more visible. There comes a time when your efficiency as an operating unit is inadequate.

You may think that this is an exaggeration so let us look at what some of the leading systems designers have had to say on the subject. I quote from Robert Boguslaw and I have checked on this quotation because I could not believe it could be serious when I first came across it. I was assured, however, that this statement was made following a series of discussions with some systems engineers at a major U.S. company.

> Our immediate concern let us remember, is the exploitation of the operating unit approach to systems design no matter what materials are used. We must take care to prevent this discussion from degenerating into the single sided analysis of the complex characteristics of one type of systems material, namely human beings. What we need is an inventory of the manner in which human behaviour can be controlled, and a description of some of the instruments which will help us achieve that control. If this provides us with sufficient handles on human materials so that we can think of them as metal parts, electrical power or chemical reactions, then we have succeeded in placing human material on the same footing as any other material and can begin to proceed with our problems of systems design. There are however, many disadvantages in the use of these human operating

units. They are somewhat fragile, they are subject to fatigue, obsolescence, disease and even death. They are frequently stupid, unreliable and limited in memory capacity. But beyond all this, they sometimes seek to design their own circuitry. This in a material is unforgivable, and any system utilising them must devise appropriate safeguards.[10]

So according to Boguslaw, that which is most precious in human beings, the ability to design their own circuitry, or to think for themselves, is now said to be an attribute which should quite deliberately be suppressed. The reason for all this is that the whole introduction of these systems is being based on the notion of Taylorism.

Frederick Winslow Taylor once said, "In my system the workman is told precisely what he is to do and how he is to do it, and any improvement he makes upon the instructions given to him is fatal to success."[11]

Taylor's philosophy is being introduced into the field of intellectual work, and in order to condition us to this subordinate role to the machine and to the control of human beings through the technology, the idea is fed out in a whole series of very interesting and subtle statements. Take this one from the Journal of Accountancy in the United States. It talks about the idiosyncracies of accountants and how you must control them when you introduce the computer. "If you have got disgruntled employees, you should not allow them to start in case they might abuse the computer." Now I would be concerned if the computer abused the employees, but the whole philosophy is that it is the machine that matters and it is the human being who has to be modified or selected for suitability.

WHY DIMINISH THE HUMAN INTELLECT?

Professor Heath of Herriot Watt University in England has talked about computers reaching an I.Q. of 120 within the next two decades. He went on to say that we will then have reached the point where we shall have to decide whether they are people or not! I don't know what he thinks of people with an I.Q. of less than 120, but is this theological debate supposed to be a serious one? Professor Heath says, "If they are people, the secular consequences are obvious. They must have the vote; switching them off

Identifying the Problem 21

would be classed as an assault and the erasure of memory as murder."[12] Gradually we are being conditioned to think that this is a valid area of discussion!

The more I look at human beings, the more impressed I become with the vast bands of intelligence they can use. We often say of a job "It's as easy as crossing a road," yet as a technologist I am ever impressed with people's ability to do just that. They go to the edge of the pavement and work out the velocity of the cars coming in both directions by calling up a massive memory bank which will establish whether it's a car or a bus because there is significance in the actual size. They then work out the rate of change of the image and from this assess the velocity. They do this for vehicles in both directions in order to assess the closing velocity between them. At the same time they are working out the width of the road and their own acceleration and peak velocity. When they decide they can go, they will just fit in between the vehicles.

The above computation is one of the simpler ones we do, but you should watch one of our skilled workers at Lucas Aerospace going through the diagnostic procedures of finding out what has gone wrong with an aircraft generator. There you see real intelligence at work. A human being using total information processing capability can bring to bear synaptic connections of 10^{14}, but the most complicated robotic device with pattern recognition capability has only about 10^3 intelligence units.

Why do we deliberately design equipment to enhance the 10^3 machine intelligence and diminish the 10^{14} intellect? Human intelligence brings with it culture, political consciousness, ideology and other aspirations. In our society these are regarded as somewhat subversive, a very good reason then to try and suppress it or eliminate it altogether, and this is the ideological assumption present all the time. (*Fig. 7*).

As designers we don't even realize we are suppressing intellects, we are so preconditioned to doing it. That is why we have this terrific drive in certain fields of artificial intelligence. Fred Margulies of the Austrian Trade Union Movement, commenting recently on this waste of human brainpower, said, "The waste is a twofold one, because we not only make no use of the resources available, we also let them perish and dwindle. Medicine has been aware of the phenomenon of atrophy for a long time. It denotes

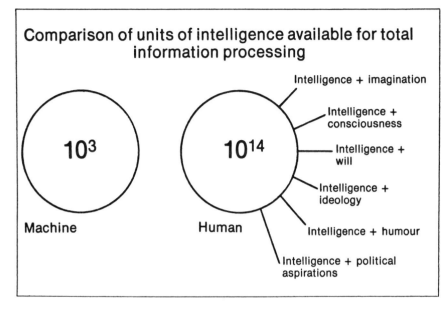

Fig 7

the shrinking of organs not in use such as muscles in plaster. More recent research of social scientists supports the hypothesis that atrophy will also apply to mental functions and abilities."

To illustrate the capabilities of human brainpower, I quote Sir William Fairbairn's definition of a millwright of 1861.

> The millwright of former days was to a great extent the sole representative of mechanical art. He was an itinerant engineer and mechanic of high reputation. He could handle the axe, the hammer and the plane with equal skill and precision; he could turn, bore or forge with the despatch of one brought up to these trades and he could set out and cut furrows of a millstone with an accuracy equal or superior to that of the miller himself. Generally, he was a fair mathematician, knew something of geometry, levelling and mensuration, and in some cases possessed a very competent knowledge of practical mathematics. He could calculate the velocities, strength and

power of machines, could draw in plan and section, and could construct buildings, conduits or water courses in all forms and under all conditions required in his professional practice. He could build bridges, cut canals and perform a variety of tasks now done by civil engineers.[13]

All the intellectual work has long since been withdrawn from the millwright's function.

TWO

The Human Machine Interaction

SQUEEZING THE MOST OUT

A few years ago, our then progressive Department of Industry produced a document that was called "Man/Machine Systems Designing" and they meant *man* as well! It was not just the generic term. The different characteristics of the human being and the machine are related in this report and the different attributes listed. Under SPEED it says, "The machine is much superior" and of the human it says, "1 second time lag." Under CONSISTENCY it says of the machine "ideal for precision" and of the human "not reliable, should be monitored by the machine." When it comes to OVERLOAD RELIABILITY it says of the machine "sudden breakdown" and of the human being "graceful degradation."

One does not need to be a sociological Einstein to work out what is going on. The people who sell this kind of equipment make it clear enough themselves. In *The Engineer* (20 June 1974) which I believe most engineers read, there was an advertisement for a package which said, "If you've got a guy who can produce drawings non stop all day, never gets tired or ill, never strikes, is happy on half pay with a photographic memory, you don't need...!" Now we know why that package is marketed. It states it clearly in the ad. *The Economist* likewise spells it out clearly enough. It points out "Robots don't strike" and it advises managements to introduce robotic equipment as a way of controlling militant workforces.

TOO OLD AT 24

Just as machines are becoming more and more specialized and dedicated, so is the human being, the "appendage" to the machine. In spite of all the talk in educational circles about wider and more generalized education, the reality is that many companies will not recruit an electronics engineer over the age of 23 and they will specify with minute precision the exact kind of engineer and specialization they want. The historical tendency is towards greater specialization in spite of all the talk about universal machines and distributed systems.

The people who interface with the machine are also required to be specialized. However, as shown on page 12 above, this is accompanied by a growing rate of knowledge obsolescence. It was recently pointed out by Eugene Wigner, when talking about the way our education system is going to meet this problem of specialization, that it is taking longer and longer to train a physicist. "It is taking so long to train him to deal with these problems that he is already too old to solve them." This is at 23 or 24 years of age!

The "peak performance age" for people of particular specializations is being worked out by a whole range of researchers. People of all different age groups sit in front of a visual display unit solving problems of growing complexity. The "response time" is plotted against the complexity of the task as shown in the graph (*Fig. 8*). It can be seen that as the tasks become more complex, the response time of the older people shows a much more marked increase.

It could be said, of course, that the older worker has a greater range of experience and knowledge and can therefore see more problems. But even if this were not the case, have we reached such a depraved stage that the natural biological process of growing old is now to be treated as a crime which must be economically

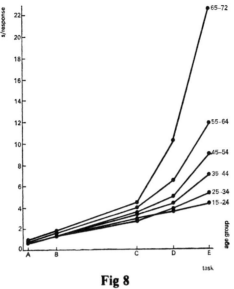

Fig 8

penalized? We design equipment to suit only the peak performance age. How many people over 40 do you see in a high pressure computerized environment, interfacing with the equipment? Yet there is nothing more natural and inevitable than growing old. "We are all born of the gravedigger's forceps" as Samuel Beckett said.

The following graph (*Fig. 9*) represents the results of experiments carried out in the U.S. A group of workers (here they are scientific workers) of various ages are given some simple but original problems to solve. The rate at which they are solved is plotted against their age and a performance curve results.

It is found that a pure mathematician reaches his or her peak performance age at about 24 or 25, a theoretical physicist at about 26 or 27 and a mechanical engineer at about 34. This latter is the most durable profession. It happens to be my own, and I am well beyond that age.

It is suggested that workers in these professions should be brought through a career pattern where they have their highest level of salary and status for a few years around their peak performance age. After that they should experience a "careers de-escalation".

If you have ever looked at a profile of a manual worker's pay related to physical prowess or work tempo, you will recognize that

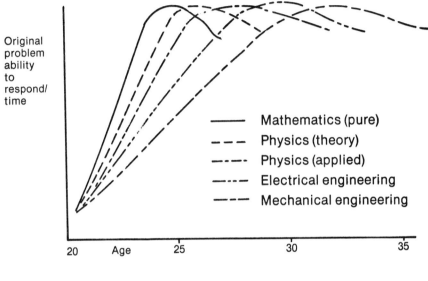

Fig 9

it is exactly the same kind of curve. In other words, it is now being repeated in the field of intellectual work. One of the justifications for this is that the increased productivity will provide the data and the time for people to be creative. The notion of increased creativity by these means seems to me to be highly questionable.

JUGGLE YOUR STANDARD BITS!

A system known appropriately as "Harness" was introduced some time ago in the field of architectural design. The idea is that you can reduce a building to a system of standardized units. In systems of this kind, all the architect can do is arrange the predetermined architectural elements around the visual display unit (V.D.U.) screen. The possibility of changing the elements becomes increasingly limited. It's somewhat like a child using a "lego set".

You can make pleasing patterns but you cannot change the form or nature of the elements themselves!

I understand from some colleagues who work in local government that if you use a system like Harness for about two years, you are then regarded by the architectural community as being de-skilled, and have great difficulty in getting jobs. This puts the architect in a similar position to the manual worker who uses a specialized lathe and cannot then get a job doing universal and more skilled work.

Likewise, the print industry is now being transformed by the use of computers. Those in the industry are being assured that it will increase their creativity as well. Apart from the jobs permanently eliminated by these new technologies and the conflict which arises (as at the *Times*), I would argue that much of the creative work within the print industry and the newspaper industry as a whole is being diminished. The new role of the journalists will be to work through a visual display unit where they prepare not merely the text, but through the computer, the typeface as well. It is suggested that since they can move sections of text around and modify sentences and paragraphs at great speed, this will increase their creativity. However, experience of these new technologies in the United States has already begun to show that it is resulting not in flexibility but in rigidity. This is because standard statements can be stored in the computer and called up when required to compose a story. This is done initially by counting through the computer the rate at which certain phrases or sentences occur. The most frequent ones are then stored and treated as optimum sentences or "preferred subroutines" which the journalist is then required to use. (We are obsessed with optimization!).

Suppose you were a reporter writing about some political activity. You would have to lead in with a sentence like "It was reported in Washington....." You couldn't say, for example, "Those idiots got it wrong again...," or some other unusual remark because it would not be an available subroutine. The individual style of a journalist which gives journalism its color and interest is gradually being diminished. There have already been complaints about some newspapers produced like this in the United States.

It is sometimes suggested that this is merely a transitional stage, a sort of industrial purgatory through which we must go on the way to a promised occupational land in which sophisticated systems and masses of data will present us with such a massive range of permutations and combinations that we can hardly fail but to be highly creative. Such a view is like that of the professor with his "contrivance" on the island of Laputa in Gulliver's Travels:

> Everybody knew how laborious the usual method is of attaining to arts and sciences, whereas by his contrivance, the most ignorant person, at a reasonable charge and with little bodily labour, may write books in philosophy, poetry, politics, law, mathematics and theology without the least assistance from genius or study.

The "contrivance" was a sort of idiot frame containing all the letters of the alphabet many times over. The pupils were trained to spin the frame continually and write down the words appearing. The logic was that if you did it often enough you could not fail to come up with something worthwhile. Just the sort of argument used in respect of computers. What a pity Gulliver's Travels is always looked upon as a children's book!

A further assumption is that "logical" information retrieval systems from which we can call up dedicated packages of knowledge, will enhance our decision making capabilities. However, as Shakel has pointed out, "Often human logic is not logical."[1] Although he was talking about voice input systems, the same may be argued for information retrieval systems. If an intelligent human being goes to a library to look up reference material, he or she will invariably be diverted off into a series of avenues which, in terms of the dedicated knowledge required, might be regarded as redundant. Yet the richness of human behavior and human intelligence comes about as a result of these wide bands of knowledge and experiences. This apparently redundant information may subsequently be vitally important on entirely different projects and in apparently unrelated fields. We have reached a serious if perhaps predictable situation, when the *Times* can announce in a headlines with apparent approval, "The Library where nobody browses and where Automation is the Chief Assistant."[2]

It is suggested that in these highly automated libraries, offices and work sitatuions, human beings will actually enjoy conversing with machines more than with people. I have even heard it said that patients prefer to converse with computers rather than with their doctors. This probably says more about the deplorable state of medicine in technologically advanced society than it does for the elegance of our computer systems design. The rich interaction which comes from people discussing work problems with each other, and the open ended intellectual cross fertilization which flows from that may well be lost, and human beings could become industrial Robinson Crusoes in an island of machines. This lack of human intercourse and social contact, together with its effect on the functioning of the brain has been discussed in a much wider context by the neurobiologist Steven Rose.[3]

It is typical of the narrow, fragmented and shortsighted view that our society takes of all productive processes that these important philosophical considerations are usually ignored.

Some design methodologists have raised these questions but the lack of any serious debate within the design community is itself indicative of the seriousness of the situation.[4] One of the founders of modern cybernetics, Norbert Wiener, once cautioned, "Although machines are theoretically subject to human criticism, such criticism may be ineffective until long after it is relevant." It is surprising that the design community, which likes to pride itself on its ability to anticipate problems and to plan ahead, shows little sign of analyzing the problems of computerization "until long after it is relevant." Indeed, in this respect, the design community is displaying in its own field, the same lack of social awareness which it displays when implementing technology in society at large.

Undoubtedly, the major part of these problems arises from the economic and social assumptions that are made when equipment of this kind is introduced. Another significant problem is the assumption that so called "scientific methods" will result inevitably in better design, when in fact there are grounds for questioning whether the design process lends itself to these would-be scientific methods.[5]

Related to this is one of the unwritten assumptions of our scientific methodology—namely, that if you cannot quantify something you pretend it doesn't actually exist. The number of complex situations which lend themselves to mathematical modelling is very small indeed. We have not yet found, nor are we likely to find, a means of mathematically modelling the human mind's imagination. Perhaps one of the positive side effects of computer aided design (C.A.D.) is that it will require us to think more fundamentally about these profound problems and to regard design as a holistic process. As Professor Lobell has put it,

> It is true that the conscious mind cannot juggle the numbers of variables necessary for a complex design problem, but this does not mean that systematic methods are the only alternative. Design is a holistic process. It is the process of putting together complex variables whose connection is not apparent by any describable system of logic. It is precisely for that reason that the most powerful logics of the deep structures of the mind, which operate free of the limitations of space, time and causality have traditionally been responsible for the most creative work in all the sciences and arts. Today it gone out of fashion to believe that these powers are in the mind.[6]

CREATIVE MINDS

It is a fact that the highly constrained and organized intellectual environment of a computerized office is remarkably at variance with the circumstances and attributes which appear to have contributed to creativity in the arts and sciences.[7][8][9] I have heard it said that if only Beethoven had had a computer available to him for generating musical combinations, the 9th symphony would have been even more beautiful. But creativity is a much more subtle process. If you look historically at creative people, they have always had an open-ended childlike curiosity. They have been highly motivated and had a sense of excitement in the work they were doing. Above all, they possessed the ability to bring an original approach to problems. They had, in other words, very fertile imaginations. It is our ability to use our imagination that distinguishes us from other animals. As Karl Marx wrote,

> A bee puts to shame many an architect in the construction of its cells; but what distinguishes the worst of architects from the best of bees is namely this. The architect will construct in his imagination that which he will ultimately erect in reality. At the end of every labour

process, we get that which existed in the consciousness of the labourer at its commencement.[10]

If we continue to design systems in the manner described earlier, we will be reducing ourselves to bee-like behavior.

It may be regarded as romantic or succumbing to mysticism to emphasize the importance of imagination and of working in a non-linear way. It is usually accepted that this type of creative approach is required in music, literature and art. It is less well recognized that this is equally important in the field of science, even in the so called harder sciences like mathematics and physics. Those who were creative recognized this themselves. Isaac Newton said, "I seem to have been only like a boy playing on the sea shore and diverting myself in now and then finding a smoother pebble or a prettier shell than ordinary, while the great ocean of truth lay all undiscovered before me."

Einstein said that "Imagination is far more important than knowledge." He went on to say, "The mere formulation of a problem is far more important than its solution which may be merely a matter of mathematical or experimental skills. To raise new questions, new possibilities and to regard old problems from a new angle requires creative imagination and marks real advances in science."

On one occasion when being pressed to say how he had arrived at the idea of relativity, he is supposed to have said, "When I was a child of 14 I asked myself what the world would look like if I rode on a beam of light." A beautiful conceptual basis for all his subsequent mathematical work.

Central to the Western scientific methodology is the notion of predictability, repeatability and quantifiability. If something is unquantifiable we have to rarify it away from reality which leads to a dangerous level of abstraction, rather like a microscopic Heisenberg principle. Such techniques may be acceptable in narrow rarified mathematical problems, but where much more complex considerations are involved, such as in the field of design, they may give rise to questionable results.

> The risk that such results may occur is inherent in the scientific method which must abstract common features away from concrete reality in order to achieve clarity and

systematisation of thought. However, within the domain of science itself, no adverse results arise because the concepts, ideas and principles are all interrelated in a carefully structured matrix of mutually supporting definitions and interpretations of experimental observation. The trouble starts when the same method is applied to situations where the numbers and complexity of factors is so great that you cannot abstract without doing some damage, and without getting an erroneous result.[11]

More recently, these questions have given rise to a serious political debate on the question of the neutrality of science and technology,[12] and there is likely to be growing concern as to the ideological assumptions built in to our scientific methodologies.[13]

THE POTENTIAL AND THE REALITY

Those who initiate scientific and technological advances, are frequently moved by the loftiest motives and display a genuine desire to improve the quality of life of those affected by their innovations. Who would doubt the motives of Pascal who, when he had designed and built the first true mechanical calculating machine in 1642 declared, "I submit to the public a small machine of my own invention, by means of which you alone may, without any effort, perform all the operations of arithmetic, and may be relieved of the work which so often fatigues your spirit when you worked with the counters and the pen."

Similarly, the motives of those innovating in the field of computer aided design are likely to be equally laudable. Professor Tom Maver and his colleagues at Strathclyde would like to see computers used to democratize the decision making process in architectural design. Arthur Llewelyn, Director of the Computer Aided Design centre at Cambridge, England, has repeatedly and so correctly asserted that computers should not be used as a means of eliminating designers and draftsmen, but rather as tools to improve their responsibility and ability to carry out creative tasks.

Regrettably, the history of scientific and technological innovation is strewn with dramatic examples which contrast the dedicated and socially desirable objectives of the academic or researcher with the cynical exploitation of their ingenuity at the level of application by the owners of the means of production. Hence we find in many fields of endeavor a significant gap between that which technology could provide (its potential) and that which it does provide (its reality).

There is a tendency, therefore, to make value judgements about given technologies based on what they might achieve rather than what they have already achieved and are likely to continue to achieve within a given economic, political and social framework. Thus, those who have no experience of C.A.D. (or very little and are therefore at the Gee Whiz stage) tend to display a more positive attitude to C.A.D. than those who have had to live with it for some time. Similarly, the genuine enthusiasm of a C.A.D. specialist on a research project in the relative monastic quiet of a university is unlikely to be shared by the designer faced with the harsh reality of its consequences in some high pressure multinational corporation. Indeed, industrial and trade union experience of computer aided design is tending to show that the thoughtless introduction of high capital equipment (computers) into the field of intellectual work (in this case design) is likely to bring in its wake many of the problems that were encountered when skilled manual work was subjected to technological change.

In academic circles, concern has recently been expressed that if we do not properly understand this historical conjuncture, we may well pursue a technological course which will permanently close off options for more humane and satisfying organizational forms in the field of intellectual work, in rather the same way we have already done in the field of craftsmanship. Failure to recognize that these options are still open to us in C.A.D. and that we still have the time and indeed the responsibility to question the linear drive forward of this technology, may well mean that we shall see growing alienation and loss of job satisfaction in engineering design. This is likely to be accompanied by the subordination of the operator (designer) to the machine (computer), with the narrow specialization of Taylorism leading to the fragmentation of design skills and a loss of panoramic view of the design activity itself. In consequence, standard routines and optimization techniques may seriously limit the creativity of the designer because the subjective value judgements would be dominated by the "objective" decision of the system. To put it another way, the quantitative elements of the design activity will be regarded as more important than the qualitative ones. There is already evidence to show that C.A.D., when introduced on the basis of so called efficiency, gives rise to a deskilling of the design function and a loss of job security—particularly for older people—which leads to structural unemployment.

To analyze why these contradictions should arise, it seems necessary to view the computer as part of a technological continuum, and its consequences as those that arise when high capital equipment is introduced into any work environment, whether it be manual or intellectual. It must also be analyzed within the economic, social and political context of the society which has given rise to the technology itself.

INDICATORS

If a comparison between design (intellectual work) and skilled craftsmanship (manual work) is really tenable, we will increasingly find strong indicators of the following:

a. The subordination of the operator (designer) to the requirements of the machine (computer) with shift work or systematic overtime to counter the increasing rate of obsolescence of the machine.
b. Emphasis upon machine centered systems rather than human centered ones.
c. Limitation of the creativity of the designer by standard routines and optimization.
d. Domination of the subjective value judgements of the designer by the "objective" decisions of the system. That is, the quantitative elements of design will be treated as more important than the qualitative ones.
e. Alienation of the designer from his or her work.
f. Abstraction of the design activity from the real world.
g. A fragmentation of design skills (overspecialization) with a loss of panoramic view together with the introduction of Taylorism and other forms of "Scientific Management", even to the extent of measuring the rate of performing intellectual work.
h. De-skilling the design function.
i. Increased work tempo as the designer is paced by the computer.
j. Increased stress, both physical and mental.

k. Loss of control over one's work environment.
l. Growing job insecurity particularly for older people.
m. Knowledge obsolescence.
n. The gradual proletarianization of the design community as a result of the tendencies indicated above, and, consequently, the considerable increase in trade union membership and industrial militancy.

C.A.D. equipment shares with all high capital equipment in a profit oriented society, the contradiction of an increasing obsolescence rate (the increasingly short life of fixed capital). Sophisticated C.A.D. equipment is now obsolete in about three years. In addition, the investment cost of the means of production (as distinct from the price of individual commodities) is ever increasing. Confronted, therefore, with equipment which is becoming obsolete literally by the minute and which requires enormous capital investment, employers will seek to exploit it 24 hours a day. This trend has long been evident on the shop floor, and the effects of shift working are already well documented. The same problems are beginning to be quite evident in the field of white collar work.[14]

Five years ago, the A.U.E.W. TASS (Amalgamated Union of Engineering Workers; Technical, Administrative and Supervisory Section) was in a major dispute with Rolls Royce, which cost the union over $500,000. The company sought, among other things, to impose on the design staff at its Bristol, England plant, the following conditions:

1. The acceptance of shiftwork in order to exploit high capital equipment.
2. The acceptance of work measurement techniques.
3. The division of work into basic elements and the setting of computerized times for these elements, such times to be compared with actual performance.

In this particular case, industrial action prevented the company from imposing these conditions. They are, however, the sort of conditions that employers will increasingly seek to impose on their white collar workers.

When staff workers, whether they be technical, administrative or clerical, work in a highly synchronized, computerized

environment, the employer will try to ensure that each element of their work is ready to feed into the process at the precise time at which it is required. A mathematician, for example, will find that s/he has to have calculations ready in the same way a Ford worker has to have the wheel ready for the car as it passes by on the production line. Consequently, we can say that the more technological change and computerization enters into white collar areas, the more workers in these areas will become proletarianized. The consequences of shift work will spread across the family, social and cultural life of the white collar worker.
that the ulcer rate of workers on a rotating shift was eight times higher than that of other workers.

> A higher proportion of night and rotating shift workers reported that they were fatigued much of the time, that their appetites were dulled and that they were often constipated.

> The most frequently mentioned difficulties in husband/wife relationships concerned the absence of the worker from the home in the evenings, sexual relations and difficulties encountered by the wife in carrying out her household duties.

> Another area of family life that seems to be adversely affected by certain kinds of shiftwork is the father/child relationship.

I quote these extracts without making any judgement about the nuclear family. I am simply indicating that the nature of technology produces effects which spread right out through the fabric of society to affect the way we live and the way we relate to other people.

The disruption of social life outside the family is also considerable. I am acquainted with a suburban estate in West London on which a number of mathematics graduates work. They used to participate in activities such as badminton, local operatics and a theatre group. When the large firm in which some of them worked introduced a computerized system it required them to work on shift. Consequently their other activities were completely disrupted.

Thus, in practice, far from humanizing the nature of work, there are grounds already for suggesting that in white collar work, high capital equipment is diminishing the quality of life of intellectual workers just as it has already done to shop floor workers.[16]

In the human/machine interaction, the human being is the dialectical opposite of the machine in that s/he is slow, inconsistent, unreliable but highly creative, whereas the machine is fast, consistent and reliable but totally non-creative. Superficially it would appear that this provides for the perfect human/machine symbiosis. In practice, the reverse is the case. As the intellectual worker tries to keep abreast of the incredible rate at which the computer can produce quantitative data, and at the same time cope with the qualitative elements, the stress upon him or her can be truly enormous. In the types of intellectual work examined by the A.U.E.W., some instances were found where the decision making rate is forced up by approximately 1900%.

A PUNISHABLE OFFENCE?

Clearly, human beings cannot stand this pace of interaction for long. Experiments have shown that the design efficiency of an engineer working at a visual display unit decreases by 30-40% in the first hour and 70-80% in the second hour.[17] Since employers, particularly in non-academic environments, will expect the equipment to be used continuously the situation can be extremely stressful. Indeed, in 1975, the International Labour Office recommended safeguards against the nervous fatigue of white collar workers. Also, an International Federation of Information Processing (IFIP) working party recently suggested that mental hazards "caused by inhumanely designed computer systems should be considered a punishable offence just as endangering the bodily safety.[18] Thus, what may be a delightfully stimulating plaything for the systems designer may be the basis for a dehumanized work environment for the user.[19]

The tendency towards automation in offices leads to a reduction in the volume of paper and to systems which are of "high information density". Micrographics systems are now commonplace peripherals to computerized systems. Complaints of eyestrain, visual discomfort, difficulties in reading and postural fatigue are now widespread. Ostberg has described some of these difficulties in an important paper which contains 84 literature references.[20]

The effects of ageing are significant for the users of these micrographic systems. Typically, for a person 16 years old with normal eyesight, about 12 dioptres of accommodation are available (the near point being 8cm) of which only one dioptre (near point at 100cm) remains at 60. Consequently, employees over the age of 50 are frequently regarded as "visually handicapped" and unsuitable for long term work with these systems. Increasingly, (particularly in Sweden) trade union and health and safety representatives are demanding that such systems should be so designed as to accommodate a wide age spectrum. These demands are part of the growing international insistence by workers on having the right to be involved in the design of their jobs, workstations and wider working environment.

A recent survey suggests that there are between 5 and 10 million visual display units currently in use. "The number is certain to increase within the next few years as the economic and other advantages of the Visual Display Unit become more apparent."[21] This same report concluded that, based on measurements and the current standards, together with the present knowledge of biological effects, the V.D.U. did not present any occupational ocular radiation hazard. Nonetheless, concern continues to grow, and a recent meeting of over 100 European experts reached the conclusion that working with V.D.U.s for 8 hours does cause fatigue, dizziness and, in extreme cases, claustrophobia. It also concluded that the operator should have frequent rests.[22]

Research programs in this field are currently in progress at the universities of Berlin, Vienna, Paris and in the U.K. at Loughborough University of Technology. Swedish trade unions have already investigated these problems and are specifying rest periods and other safeguards.[23] Far more important would be to insist upon the redesign of the equipment.

For quite some time, continental trade unions have been specifying and fighting for rest periods for the operators of this kind of equipment.[24] Trade union concern is now growing rapidly, and many trade unions are now producing check lists for the installation and use of V.D.U.s.[25] These are based on the recommendations of the International Federation of Commercial, Clerical and Technical Employees. They recommend regular eye checks at six monthly intervals, specify frequency rates, luminance of the characters on the screen, character size, shape form and height ratios. They also cover matters such as ambient lighting.

Far less research is devoted to the more subjective concerns of workers using V.D.U.s. Journalists, for example, are already complaining that the equipment gives them a feeling of isolation.[26] Managers, using an Electronic Office system established by Citibank in New York, regarded the software as "hostile" when using advanced management workstations. When a redesign of the workstation was undertaken, the philosophy was to keep existing procedures as they were and to build an electrical analogue of them. In this way, it is said, the receptivity of the users was greatly increased.[27]

More dramatic reactions by organized workers are already being reported. Thus in Norway, workers at the Norwegian Electricity Board made it quite clear to the management that they would ban a range of terminals the company intended purchasing because these could only be operated in a mode which was "unidirectional", and hence not really responsive to the human being. Such a system, they pointed out, would be inherently undemocratic and was therefore unacceptable.

The employer was then required to purchase a different range of terminals as a result of the direct industrial and collective strength of these workers. It is quite conceivable that these workers would in any case have had a constitutional right to insist on such changes. A recent Act in Norway requires employers to provide "sound contract conditions and meaningful occupation for the individual employee," "the individual employee's opportunity for self determination" and "each employer shall cooperate to provide a fully satisfactory working environment for all employees at the workplace."[28]

TAYLOR'S SCIENTIFIC MANAGEMENT

Central to the dehumanization of work in the intellectual field, just as in the field of manual work, is the fragmentation of work into narrow, alienated tasks, each minutely timed. To reduce the worker to a blind unthinking appendage of the machine is the very essence of "Scientific Management". Paradoxically, Taylor's Scientific Management applied to the shop floor initially increased the intellectual activity of the staff in the offices. In his book *Shop Management* Taylor himself explained that his system

> ...is aimed at establishing a clear cut and novel division of mental and manual work throughout the workshops. It is based upon the precise time and motion study of each worker's job in isolation, and relegates the entire mental parts of the task in hand to the managerial staff.

Timely warnings of these dangers came from 19th century writers. "To subdivide a man is to assassinate him. The subdivision of labour is the assassination of a people."[29]

The notion of the division of labor and the efficiency which is said to flow from it is normally associated with Adam Smith.[30] In fact, Adam Smith's specific arguments were anticipated by Henry Martyn almost a century earlier.[31] However, the basic notion of the division of labor is so intertwined with Western philosophy and scientific methodology that it is identifiable as far back as Plato when he argues for political institutions of the republic on the basis of the virtues of specialization in the economic sphere.

The division of labor and fragmentation of skills is of course absolutely rational if you regard people as mere units of production and are concerned solely with the maximization of the profit you extract from them. Indeed, viewed from that premise, it is not merely rational but also scientific. The scale and nature of the deskilling which accompanies this scientific management has been graphically described by Braverman.[32] This deskilling stretches right through the intellectual field.

One researcher who has examined the effects of automation in Swedish Banks states, "Increased automation converted tellers, who were in effect mini-bankers, into automatons."[33]

At the other end of the creative spectrum, those engaged in architecture, "the queen of arts rather than the father of technology," are likewise deskilled. Architects, using interactive graphic systems with packages such as Harness are confronted with a systematized approach to the design of buildings in which they are required to use a number of predetermined architectural elements. They are thus reduced to using a sophisticated lego set and their creativity is limited to choosing how the elements might be disposed by shifting them around a V.D.U. screen. The elements themselves cannot be changed. They have been optimized!

It might be argued in defense of these developments, that at least in the "occupational growth areas" associated with computing, those workers concerned with issuing instructions to the machines will be undertaking work of growing skill and creativity. To suggest this would be to fail completely to understand the historical tendency to deskill *all* work. Programming is itself being reduced to routines and "the deskiller is deskilled" as structural programming breaks with the universal (if short) tradition of idiosyncratic software production.[34]

The use of this Scientific Management has seen the fragmentation of work occurring through the spectrum of workshop activity engulfing even the most creative and satisfying manual jobs (such as toolmaking). Throughout this period, most industrial laboratories, design offices and administrative centers were the sanctuaries of the conceptual planning and administrative aspects of work. In these areas, one spur to output was a dedication to the task in hand, an interest in it, and the satisfaction of dealing with a job from start to finish. Some observers, including the author, cautioned that the situation would soon be brought to an end as the monopolies, in their quest for increased profits, bring their "rational and scientific" methods into these more self organizing and comparatively easygoing fields. The objective circumstances for this were already set when in some industries 50 or 60% of those employed are scientific, technical and managerial staff.

It was evident that the more science ceased to be an amateur gentleman's affair and was integrated into the productive processes, the more scientists and technologists would become part of the workforce itself. Indeed, it was even suggested that as high capital equipment such as computers became available to scientists and technologists, they would be paced by the machine. Eventually, their intellectual activity would be divided into routine tasks and work study would be used to set precise times for its synchronization with the rest of the "rational procedure".

Those scientists and technologists, particularly in the computer field, who look upon this view with derision, would be well advised to recall what the father of their industry, Charles Babbage, had to say on the matter. Even in the 1830s he anticipated Taylorism in the field of intellectual work. In a chapter entitled "On the Division of Manual Labour" his message is clear:

> We may have already mentioned what may perhaps appear paradoxical to some of our readers, that the division of labour can be applied with equal success to mental as well as mechanical operations, and that it ensures in both the same economy of time.[35]

THERBLIGS AND YALCS

In spite of these warnings and in spite of strikes by some white-collar unions against the use of the stopwatch in offices, these predictions were for the most part treated either as the scaremongering of slick trade union leaders keen on increasing their flock, or as plain absurdity. "That will be the day when someone tries to measure *my* intellectual activity" was a frequent reaction. Unfortunately, the day may be much closer than many would like to believe. In June 1974 there appeared in the publication *Workstudy*, "A Classification and Terminology of Mental Work." It suggests that much "progress" has been made in this direction. Having identified the hierarchy of physical work—i.e., job, operation, element, therblig, it states,

The Human Machine Interaction 45

The first three of these are general concepts—i.e., they can be applied equally well to physical or mental work. The last term, therblig, is specific to physical work. All elements of physical work consist of a small number of basic physical motions first codified by Gilbreth (Therblig is an anagram of Gilbreth) and later amended by the American Society of Mechanical Engineers and in the British Standard Glossary. The logical pattern would be complete if a similar breakdown of elements into basic mental motions—or Yalcs—were available. (Yalc is named after Clay).

The paper describes how to classify yalcs into input, output and processing yalcs, and also how each of these can be subdivided into basic mental operations. It even draws a distinction between "seeing" or the passive reception of visual signals, and "looking", i.e., their active reception. Similarly it distinguishes between "hearing" or the passive reception of audio signals and "listening", i.e., their active reception. The paper implies that these techniques will be used in the more simple aspects of mental work. However, it concludes by saying,

> We have tried to show that mental work is a valid and practical field for the application of workstudy; that basic mental motions exist and can be identified and classified in a meaningful way provided one does not trespass too far into the more complex mental routines and processes. A set of basic mental motions have been identified, named, described and coded as a basis for future work measurement research leading to the compilation of standard times. There are good prospects that such times could play a valuable part in workstudy projects.

It is clear, however, that these techniques *will* "trespass too far" into the more complex mental routines and processes, just as they have in the case of highly creative manual work. Whether one regards this type of research as pseudo scientific or not, there can be little doubt about how it will be deployed.

The employers of scientific, technical and administrative staff, including some forms of managerial staff, will see it as a powerful form of psychological intimidation to mold their intellectual workers to the "mental production line". It is perhaps a recognition of this tactical importance which prompted Howard C. Carlson, a psychologist employed by General Motors, to say, "The Computer may be to middle management what the assembly line is to the hourly paid worker."[36]

"OBJECTIVE" SCIENTIFIC DECISION MAKING

The computer is not only used as a Trojan Horse for Taylorism in the fields of management and scientific work. Even the university is no longer a sanctuary for non-alienated work. Those academics engaged in the physical and pure sciences will be pleased to learn that these important issues of efficiency and optimization will not be left to the subjective ramblings of the sociologist, or the tainted ideology of the political economist. The full analytical power and neutrality of real science and the penetrating logic of mathematical method have been brought to bear. They will undoubtedly produce a completely "objective" solution to the problem of university efficiency. For example, the notion of utilizing factory models to optimize university and polytechnic productivity has been seriously proposed. We now have the rather ironic development where some of those university people who helped to develop the scientific management productions systems which made work so grotesque for the shop floor, may soon be the victims of their own repressive techniques. An article entitled "College of Business Administration as a Production System"[37] is symptomatic of a general tendency. This article employs the terminology used to describe academic features and activities in the form of a factory model. It is strongly indicative of the underlying philosophy.

Thus the recruitment of students is referred to as "material procurement," recruiting of faculty as "resource planning and development," faculty research and study as "supplies procurement," instructional methods planning as "process planning," examinations and award of credits as "quality control," instructor evaluation as "resource maintenance" and graduation as "delivery." The professors and lecturers are of course "operators" and presumably, as on the factory floor only the effective operators will be tolerated (effective for what, and for whom, we may ask).

The administrators' definition of effectiveness and competence makes it highly likely that many of the cherished academic freedoms of the university, whether real or imaginary, will be dented. In the not too distant future, many faculty members may well find themselves subordinated to the process in the interests of efficiency, as are workers on the shop floor. To get down to the real "science" of it we can look at the proposals of Geoffrion, Dyer and Freiberg in "An Interactive Approach to Multicriterion Optimisation, with an Application to the Operation of an Academic Department."[38] They use the well known Frank-Wolfe algorithm and suggest that the multicriterion problem be reduced to the following expression:

Maximise $U [f_1(x), f_2(x), ... f_r(x)]$, subject to xeE where $f_1...f_r$ are r distinct criterion functions of the decision vector x, X is the constrained set of feasible decisions, and U is the decision maker's overall preference function defined on the values of criteria.

Taking a specific department as an example, they define six criteria for it. The first three are the number of course sections offered by the department at graduate, lower division undergraduate and upper division undergraduate levels. Criterion four is the amount of teaching assistant time used for the support of classroom instruction by the faculty. The fifth criterion is the regular faculty effort devoted to major departmental duties measured in equivalent course sections. Finally, criterion six is the regular faculty effort devoted to additional activities such as research, student counselling and minor administrative tasks, again measured in course sections.

Terms such as "teaching time," "teaching loads" and "faculty effort" are used throughout. This will mean that whoever makes a decision about criteria weight must have very precise times for the different functions, and thus the basis is clearly set for work measurement not unlike that on the shop floor. The justification will undoubtedly be that such times are necessary to be fed into the computerized model for objective assessment.

However, despite the veneer of mathematical objectivity, it is the subjective judgement of the so called decision maker that determines the key U function. This decision maker will be an administrator, not the academic staff themselves who will consequently experience a loss of control over their work environment. If, for example, a faculty member is informed via the computer that he or she is taking too long on teaching or spending too much time in research, or has been rendered superfluous as a result of an optimization routine (a function which mathematically illiterate workers call "the sack"), it will be worth recalling that it is the U function that predominates!

In furtherance of this efficiency, a comprehensive faculty activity analysis was prepared and developed by the University of Washington.[39] The percentage time devoted to each faculty activity is requested. All university activities, whether regular or irregular, are refined and coded. For example, code 501 (unscheduled teaching) includes thesis committee participation, discussion with colleagues about teaching, guest lecturing in other faculty members' courses and giving seminars within the institution. Each activity is specified very precisely as it might be in a factory situation. Under the code "Specific Scholarly Project" are listed: departmental research, sponsored research, writing or developing research proposals, writing books and articles and many others. Even one's own reading activities are included. Thus, under "General Scholarly Projects" we find: reading articles and books related to the profession, attending professional meetings, research related discussions with colleagues, and reviewing colleagues' research work.

There are some academics who hope that in projects of this kind educational requirements will outweigh mere productivity-oriented ones, but many feel the outcome will be a shrinking of facilities, as in the City University of New York where 700 faculty members were sacked.[40] In the United States, these programs are in fact spreading rapidly as indicated by the scale of recent grants. In the California State University and College System a "Center for Professional Development" was set up with a grant of $341,261 from the Fund for the Improvement of Post Secondary Education in Washington. There is no doubt what the term "Improvement" is intended to imply.

This increased productivity, however, could have consequences much more widespread and subtle than the obvious ones of increased work tempo, loss of control, job insecurity and even redundancy. The impact this will have on the creativity of those involved is likely to be significant, for central to all optimization procedures of this kind is the notion of specific goal objectives.

A vivid example of the need to avoid such an overconstrained work environment was the design recently of E.M.I.'s computer controlled brain and body X-ray scanner. In his evidence to the Select Committee on Science and Technology, Dr. John Powell, E.M.I.'s Managing Director, pointed out that the scanner was developed using unallocated funds as a by-product of work on optical character recognition, Dr. Powell stated that had its inventor "been constrained to follow a set objective on contract research funded by an operating division, he might have just produced another optical character recognition machine."

The inventor himself, Dr. Godfrey Hounsfield, who received the 1979 Nobel Prize for medicine as a result of his work on the scanner, said about his habit of going for long walks, "It is a time when things come to one I find. The seeds of what happened came on a ramble." He said also, "I still feel quite a lift when I find that the machine is doing good."[41]

THREE

Political Implications of New Technology

Scientific and technical advance, in spite of its liberatory potential, brings also in its wake powerful tendencies of control and authoritarian organizational forms. Indeed, it has been suggested that "control" has been as much a stimulus to technological change as has "productivity." Some researchers pointed out as early as the 50s and 60s that computers increase the authoritarian control which an employer has over his employees and strengthens the hand of those who support a tougher attitude to employees.[2]

The process is succinctly described in the magazine *Realtime* (issue 6, 1973) by a writer fresh from an I.B.M. customer training course.

> Now an operating system is a piece of software functionally designed to do most efficiently a particular job—or is it? It gradually dawned on me that some rather obnoxious cultural assumptions have

been imported lock, stock and barrel into I.B.M. software. Insidious, persuasive assumptions which appear to be a natural product of logic—but are they?

The whole thing is a complete totalitarian hierarchy. The operating system runs the computer installation. The chief and most privileged element is the "Supervisor." Always resident in the most senior position in the main storage, it controls through its minions, the entire operation. Subservient to the Supervisor is the bureaucratic machinery—job management routines, task management, input/output scheduling, spares management and so on. The whole thing is thought out as a rigidly controlled, centralised hierarchy, and as machines get bigger and more powerful, so the operating system grows and takes more powers.

One lecturer soared into eloquence in comparing the various parts of the operating system to the directors, top management, middle management, shop foremen and ordinary pleb workers of a typical commercial company. In fact, the whole of I.B.M. terminology is riddled with class expressions such as master files, slave cylinders, high and low level languages, controllers, scheduler, monitor.

The same writer then generalized some of the contradictions of centralized operating systems. These coincided closely with my own findings when I investigated the contradictions in the specific field of Computer Aided Design.

The drawbacks of the centralised operating system are many. It is a constraining and conservative force. A set of possibilities for the computer system is chosen at a point in time and a change involves regeneration of the system. It imposes conformity on programming methods and thought. Another amazingly apt quote from an I.B.M. lecturer was "always stick to what the system provides, otherwise you may get into trouble." It mystifies the computer system by putting its most vital functions into a software package which is beyond the control and comprehension of the applications engineer, thus introducing even into the exclusive province of Data Processing, the division between software experts and other programmers, and reinforcing the idea that we do not really control the tools we use, but can only do something if the operating system lets you—a phrase which I am sure many of us have used. The system which results seems absurdly top heavy and complex. The need to have everything centrally controlled seems to impose an enormous strain.

MALE/FEMALE VALUES

The introduction of a computerized system is frequently used as a smokescreen to introduce another management control weapon—job evaluation. Pseudo-scientific reasons are given for fragmenting jobs and slotting the subdivided function into a low level of the system's hierarchy with correspondingly low wages for "appropriate" job grades. My experience of this in industry tends to show that it is frequently used to consolidate the unequal pay and opportunities for women. This is done either by implying, or by ensuring by structural means and recruitment, that the fragmented functions are women's work. This of course can no longer be stated openly since there is the sex discrimination legislation to watch out for, but it still happens that women are recruited for the grotesque preparation of the cards, for example, whereas the higher status jobs are offered to men.

There is, of course, no such thing as "women's work" any more than there is women's mathematics, women's physics, women's literature or women's music. There is only work; the means by which employers extract profits from all of us but higher profits from women. Thus a contradiction exists in that although scientific and technological progress could provide the objective circumstances for greater equality between the sexes in the productive process, in our profit oriented society the reverse will frequently be the case.

Women are going to have to fight, not only the traditional forms of discrimination, but much more sophisticated and scientifically structured ones. There is little indication, even in 1980, that the unions catering for such workers have really understood the nature and the scale of this problem.

When we looked over some past issues of the computing magazines covering a period of six months, 82% of the advertizements having one person in a photograph with the equipment, showed a woman in some kind of absurd posture which was in no way related to the use of the equipment. There is a continual projection of the view, even in the most serious of journals, that women are to be regarded as ridiculous playthings, just draped around the place for decoration.

Not only that, but those who read these journals often do not notice the built in assumption unless it is pointed out to them. They are conditioned to accept the presence of women in the servicing role and the absence of women in the organizing role as being quite normal. Even women themselves quite often see nothing unusual in this situation.

One of the major problems with Western science and technology is that it has the historically determined male values built into it. If we reflect upon the science that underlies most of new technology, we will see that it has three main elements: predictability, repeatability and mathematical quantifiability. As such, it puts the objective before the subjective, the quantitative before the qualitative, the digital before the analogical, discipline before freedom and the product before the producer.

In Western scientific methodology, which is based on the natural sciences, relationships are mathematically quantifiable. There is a tendency, therefore, to suggest that if you cannot quantify something, it really doesn't exist. Thus one sneers at the intuitive, tacit knowledge, practical experience and sense of feel.

This is not without its political significance; if the mass of ordinary people are incapable of providing scientific reasons for their judgments, which are based on actual experience of the real world, ruling elites can then bludgeon people's common sense into silence with quantification. This has caused the brilliant French mathematician, Professor Jean-Louis Rigal, to observe, "Quantification is the ultimate form of fascism."

At the other end of the spectrum, many of the most sensitive, creative and talented young people have decided not to study science and technology because they see it as a repressive activity. They view it as an evil and totalitarian subject devoid of those attributes which could make it amenable to the human spirit. Indeed, the writer and sociologist, D. Yankelovich, has stated in his book, *The Changing Values on the Campus*, that the student population in the U.S. included the following words among those terms it rejects as "bad": verification, facts, technology, statistical controls, programming, calculate, objectify, detachment. Not surprisingly, many of these students opt for the arts or the social sciences where they feel (sometimes mistakenly) that more opportunity will exist for humanistic concerns.

What we are beginning to witness on many fronts is a questioning of the value systems which advise our science and technology. The neutrality of science is no longer taken for granted, and the issues involved transcend those of a simplistic use/abuse model and lead to the deeper considerations of the nature of the scientific processes themselves. Science performed within a particular social order reflects the norms and ideology of that social order. Viewed thus, science ceases to be regarded as autonomous and is seen as part of an interacting system in which internalized ideological assumptions help to determine and form the very experimental designs and theories of the scientists themselves.

It reflects the economic base and power relations of the society which has given rise to it and displays predominantly the value system of the white male capitalist warrior, admired for his strength and speed in eliminating the weak, conquering competitors and ruling over vast armies of men who obey his every instruction. He makes decisions which are logical and rational, and which will lead to victory.

Technological change is starved of the so called female values such as intuition, subjectivity, tenacity and compassion. It would be an enormous political and philosophical contribution to society if more women were to come into science and technology, not as imitation men or honorary males, but to challenge the "male" values which have distorted "science" for too long. It would be a contribution, not only to women's liberation and that of humanity as a whole, but also toward science, which would become more caring, liberatory, socially relevant and natural.

Until then, if a feminist scientist were to say, "The really valuable factor is intuition," she would be laughed out of the laboratory by male colleagues, who wouldn't even realize that she was quoting Einstein!

There are additional and substantial concerns which must surround the use of computers. If human beings increasingly work with models of reality rather than with reality itself and are thereby denied the precious learning process which flows from it and the accumulation of tacit knowledge, the problems are likely to be significant and have been discussed by writers of widely varying "political stances."[3][4][5]

Many designers fear to discuss these concerns because they may be accused of being "unscientific." There is no suggestion in this line of argument that one should abandon the "scientific method"; rather, we should understand that this method is merely complementary to experience and should not override it, and that experience includes "experience of self as a specifically and differentially existing part of the universe of reality."[6] Such a view would help us to escape from the dangers of scientism which, as was once suggested, may be nothing more than a Euro-American disease.[7]

IS SCIENCE NEUTRAL?

Marxist critics of capitalist society have tended to concentrate, at least since the turn of the century, on the contradictions of distribution. This they have done at the expense of a thorough going analysis of the contradictions of production within technologically advanced society.

This imbalance can hardly be attributed to a onesidedness on Marx's own part. Central to volume I of *Capital* is the nature of the labor process and a "critical analysis of capitalist production." In this, Marx demonstrates that with the accumulation of capital—the principal motivating force—the processes of production are incessantly transformed. For those who work, whether by hand or brain, this transformation shows itself as a continuous technological change within the labor process of each branch of industry, and secondly, as dramatic redistributions of labor among occupations and industries.

That the overall development of production since then should accord so closely with Marx's analysis is a remarkable tribute to his work, especially if one bears in mind the sparsity of occupations and industries then, compared with their proliferation today. Whether this Marxist analysis will be equally consistent and valid when applied to the science based industries which have emerged since the second world war is now a matter of considerable discussion. With the integration of science into the "productive forces" this question is one of growing significance. In some large multinational corporations 50% or more of all those employed are scientific, technical or administrative "workers." This has begun to pose, in a very practical way, the relationships between science as presently practiced and society.

THE USE/ABUSE MODEL

Up to the mid-sixties, there hardly seemed any useful purpose in raising this question. At that time, there was hardly a chink in the Bernalian analysis of 20th century science. In this analysis, science, although it was integral to capitalism, was ultimately in contradiction with it. Capitalism, it was felt, continuously frustrated the potential of science for human good. Thus, the problems thrown up by the application of science and technology were viewed simply as capitalism's misuse of their potential. The contradictions between science and capitalism were viewed as the inability of capitalism to invest adequately, to plan for science, and to provide a rational framework for its widespread application in the elimination of disease, poverty and toil

The forces of production, in particular science and technology, were viewed as ideologically neutral, and it was considered that the development of these forces was inherently positive and progressive. It was held that the more these productive forces—technology, science, human skill, knowledge and abundant "dead labor" (fixed capital)—developed under capitalism, the easier the transition to socialism would be. Further, science is rational and could therefore be counterposed against irrationality and suspicion.

Science had, after all, through the Galilean revolution, destroyed the earth-centered model of the universe, and through Darwin had made redundant earlier ideas of the creation of life and of humanity. Science viewed thus appeared as critical knowledge, liberating humanity from the bondage of superstition. Superstition, which was elaborated into the system of religion, had acted as a key ideological prop of the outgoing social order.[8] The past few years have seen a growing questioning of this rather mechanistic interpretation of the Marxian thesis. There is now a growing realization that science has embodied within it many of the ideological assumptions of the society which has given rise to it. This in turn has resulted in a questioning of the neutrality of science as presently practiced in our society. The debate on this issue is likely to be one of major political significance. The question extends far beyond that of scientific abuses to the deeper considerations of the nature of the scientific process itself. Science done within a particular social order reflects the norms and ideology of that social order. Science ceases to be seen as autonomous, but instead as part of an interacting system in which internalized ideological assumptions help to determine the very experimental designs and theories of scientists themselves.[9]

Failure to deal with these questions will mean that the anti-science movement of the 1970s, which had its antecedents in the anti-culture movement of the 1960s, will not be developed beyond its initial and partly negative premise. In this, science is viewed as evil, totalitarian and devoid of those attributes which make it amenable to the "human spirit." This total rejection, now common among many young people, can, if properly handled, be elevated into a much more mature questioning of the fundamental nature of science and technology as practiced in Western society.

It is significant that those working in the scientific field are themselves beginning to discuss this question. Professor Silver, for example, says, that there are risks

> in the scientific method, which may abstract common features away from concrete reality in order to achieve clarity and systematisation of thought. However, within the domain of science itself, no adverse effects arise because the concepts, ideas and principles are all interrelated in a carefully structured matrix of mutually supporting definitions and interpretions of experimental observation. The trouble starts when the same method is applied to situations where the number and complexity of factors is so great that you cannot abstract without doing some damage, and without getting an erroneous result.[10]

Those working in the field of cybernetics have also expressed their concern about this misuse of "science." "There is no doubt that a very important influence nowadays is a revised reductionism within the theory of cybernetics. It reduces processes and complex objectives to black boxes and dynamic control systems. Not only in the natural sciences, but also in the social sciences."[11] (My translation) My view of knowledge implies a society which has a social structure capable of nurturing the co-existence of the subjective and the objective, tacit knowledge based on contact with the physical world, and abstracted scientific knowledge. In a word, a society which would again link hand and brain and permit people to be full human beings. This will mean challenging the fundamental assumptions of our present society, and indeed the assumptions of societies such as exist in the so called socialist countries. One of the important factors now molding the social forces to give rise to such a challenge is the contradictions of science and technology experienced by an ever increasing section of the population.

The elitist right of the scientific worker or the researcher to give vent to his or her creativity will now be increasingly curbed by the system as it seeks to control more and more human behavior in all its aspects. This is part of the general attempt of the small elite who control society to gain complete control over all those who work, whether by hand or by brain, and use scientific management and notions of efficiency as a vehicle for doing so. It will be seen then that the organization of work, as well as the means of designing both jobs and the machines and computers necessary to perform them, embodies profound ideological assumptions. So by regarding science and technology as neutral, we have

> failed to recognise as anti-human, and consequently to oppose the effects of values built into the apparatus, instruments and machines of their capitalist technological system. So, machines have played the part of a Trojan Horse in their relation to the Labour Movement. Productivity becomes more important than fraternity. Discipline outweighs freedom. The product is in fact more important than the producer, even in countries struggling for socialism.[12]

It has been suggested that by ignoring these considerations the Soviet Union was laying the basis for the present situation in which it would be hard to argue that a worker there enjoys a sense of fulfillment through his or her work envisioned by the early Marxists. It may well be, that instead of developing entirely different forms of science and technology and merely trying to adapt those developed in the capitalist societies to their society, the Soviet Union has made a profound error. Indeed, the development in that country must find part of its origins in the attitude of Lenin to Taylorism; which, he said,

> Like all capitalist progress is a combination of the refined brutality of bourgeois exploitation, and a number of the greatest scientific achievements in the field of analyzing mechanical motions during work, the elimination of superfluous and awkward motions, the elaboration of the correct methods of work, the introduction of the best system of accounting and control, etc., the Soviet Republic must at all costs

adapt all that is valuable in the achievement of Science and Technology in this field. The possibility of building socialism depends exactly on our success in combining the Soviet Power and the Soviet Organisation of Industry with the up to date achievements of capitalism. We must organise in Russia the study and teaching of the Taylor system, and systematically try it out and adapt it to our ends.[14]

Socialism, if it is to mean anything, must mean more freedom rather than less. If a worker is constrained through Taylorism at the point of production, it is inconceivable that he or she will develop the self confidence and the range of skills, abilities and talents which will make it possible to play a vigorous and creative part in society as a whole.

So it is, that in the technologically advanced nations, there are now beginning to emerge a range of contradictions that will necessitate a radical examination of how we use science and technology, and how knowledge should be applied in society to extend human freedom and development.

TECHNOLOGICAL CHANGE AND PROLETARIANIZATION

The emergence of fixed capital as a dominant feature in the productive process means that the organic composition of capital is increased and industry becomes capital intensive rather than labor intensive. Human beings are increasingly replaced by machines. This in itself increases the instability of capitalism; on the one hand capitalism uses the quantity of working time as the sole determining element, yet at the same time continuously reduces the amount of direct labor involved in the production of commodities. At an industrial level, literally millions of workers lose their jobs and millions more suffer the nagging insecurity of the threat of redundancy. An important new political element in this is the class composition of those being made redundant. Just as the use of high capital equipment has spread out into white collar and professional fields, so also has the consequences of this use of high capital equipment done likewise. Scientists, technologists, professional workers and clerical workers all now experience unemployment in a manner that only manual workers did in the past. Verbal niceties are used to disguise their common plight. A large West London engineering organization declared its scientists and technologists "technologically displaced," its clerical and administrative workers "surplus to requirements" and its manual workers "redundant". In fact, they had all got the good old fashioned sack! In spite of different social, cultural and educational backgrounds, they all had a common interest in fighting the closure of that plant, and they did. Scientists and technologists paraded around the factory carrying banners demanding "the right to work" in a struggle that would have been inconceivable a mere ten years ago. Technological change was indeed proletarianizing them. In consequence of the massive and synchronized scale of production which modern technology requires, redundancies can affect whole communities. During a recent recession in the U.S. aircraft industry, a union banner read:

Last out of Seattle please put the lights out.

Because of this change in the organic compositon of capital, society is gradually being conditioned to accept the idea of a permanent pool of unemployed persons. Thus we find in the United States during the last half of the 1970s over 7 million people have been permanently out of work, and that's based on the "official" unemployment rate!

We have witnessed in England the large scale unemployment of recent years. Unemployment is considerable in Italy, and even in the West German miracle there are sections of workers — particularly over the age of fifty—who are now experiencing long terms of unemployment. This unemployment itself creates contradictions for the ruling class. It does so because people have a dual role in society, that of producers and consumers. When you deny them the right to produce, you also limit their consumption power. In the late 60s and early 70s efforts were made to restructure the Social Services to maintain the balance between unemployment and the purchasing power of the community. However, now due to the growing fiscal crisis of the state, Social Services are themselves being drastically cut in both England and the U.S. In the early 1960s

President Kennedy spoke of a "tolerable level of unemployment." In Britain in the the 1960s Harold Wilson, having fueled the fires of industry with the taxpayers' money to create the "white heat of technological change," spoke in a typical double negative of a "not unacceptable level of unemployment."

A remarkable statement for a so-called socialist Prime Minister! The net result is that there is on the one hand an increased work tempo for those in industry, while on the other hand there is a growing pool of unemployed, with all the degradation that implies. Nor is there any indication that the actual working week has been reduced during this period. Indeed, in spite of all the technological change since the Second World War, the actual working week in Britain for those who have the jobs is now longer than it was in 1946. Yet the relentless drive goes on to design machines and equipment which will replace workers. Those involved in such work seldom question the nature of the process in which they are engaged. Why, for example, the frantic efforts to design robots with pattern recognition intelligence when we have seven hundred thousand people in the dole queue in Britain whose pattern recognition intelligence is infinitely greater than anything yet conceived even at a theoretical level?

That this should be so reveals the extent to which we have been conditioned by the criteria of the market economy. Thus, we see the freeing of capital as an asset, and the freeing of people as a liability. In doing this we ignore our most precious asset — people, with their skill, ingenuity and creativity. In the defense and aerospace industries we have some of the most highly skilled and talented workers in this country. Yet, like the ruling class we have thought of capital first and people last, and ignored the incredible contribution which their skill and ability could make to the well being of the people of this country.

Confronted with these contradictions, the bleating and whimpering of the European and U.S. trade union bureaucracies (to be contrasted with the creative Luddism of a least some sections of the Australian movement) failed to disguise the reality that they have no independent view of how science and technology should develop. Indeed, when they are not demanding more investment in the same forms of technology that have given rise to the problems in the first place, they are making minor, pathetic, window dressing modifications to the proposals of the vast multinational corporations. A typical trade union response is illustrated in the diagram. *See Fig. 1.*

Given the gradual incorporation of the trade unions through indicative planning, and in Britain — quangos, this is perhaps not so surprising. What is, however, disconcerting, is the total disarray and confusion of the Marxist Left as the political pigeons of blind unthinking technological optimism come home to roost with a vengeance.

"USING" PEOPLE

The system seeks in every way to break down workers' resistance to being sacked. One of the sophisticated devices was the Redundancy Payments Act* under the Labor Government. Practical experience of Trade Unions in Britain demonstrates that the lump sums involved broke up the solidarity at a number of plants where struggle was taking place against a closure.

A much more insidious device is to condition the workers into believing that it is their own fault that they are out of work, and that they are, in fact, unemployable. This technique is already widespread in the United States where it is asserted that certain workers do not have the intelligence and the training to be employed in modern technological society. This argument is particularly used against black workers, Puerto Ricans and poor whites. There is perhaps here fertile ground for some of the "objective research" of Jensen and Eysenk.

The concept of a permanent pool of unemployed persons as a result of technological change, also brings with it the danger that those unemployed would be used as a disciplining force against those still in work. It undoubtedly provides a useful pool from which the army and police force can draw, and during the recent redundancies in Britain, a considerable

---

*The Redundancy Payments Act provided for a lump sum payment by the state and employers to those made redundant. The payment was based on the length of service in the job declared redundant.

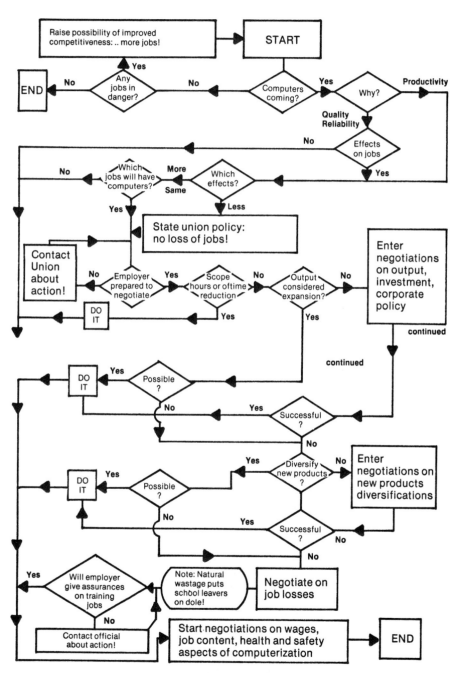

Fig 1

number of fired workers from the North East were recruited into the army and then used against workers in Northern Ireland. Coupled with the introduction of this high capital equipment is usually a restructuring known as "rationalization". The epitome of this in Britain is the G.E.C. complex with Arnold Weinstock at its head.

In 1968, this organization employed 260,000 workers and made a profit of over $150 million. In consequence of quite brutal layoffs, the company's work force was reduced to 200,000 yet profits went up to over $210 million. These are the kind of people who are introducing high capital equipment, and they make their attitude to human beings absolutely clear. It is certainly profits first and people last! One quotes Arnold Weinstock not because he is particularly heinous (he is, in fact, extremely honest, direct and frank) but because he is prepared to say what others think. He said on one occasion "people are like elastic, the more work you give them, the more they stretch." We know, however, that when people are stretched beyond a limit they break. My union has identified a department in a West London engineering company where the design staff was reduced from 35 to 17 and there were six nervous breakdowns in eighteen months. Yet people like Weinstock are held up as a glowing example to all aspiring managers. One of his own senior managers once boasted that "he takes people and squeezes them till the pips squeek." I think it is a pretty sick and decaying society that will boast of this kind of behavior.

Most industrial processes, however capital intensive they might be, still require human beings in the total system. Since a highly mechanized or automated plant frequently is capable of operating at very high speeds, employers view the comparative slowness of the human being in his interaction with the machinery as a bottleneck in the overall system. In consequence of this, pay structures and productivity deals are arranged to ensure that the workers operate at an even faster tempo.

FRANTIC TEMPO

In many instances the work tempo is literally frantic. In one automobile factory in the Midlands in Britain they reckon that they "burn a man up" on the main production line in ten years. They recently tried to get our union to agree that nobody would be recruited for this type of work over the age of 30.

For the employer it is like having a horse or dog. If you must have one at all then you have a young one so that it is energetic and frisky enough to do your bidding all the time. So totally does the employer seek to subordinate the worker to production, that he asserts that the worker's every minute and every movement "belong" to him, the employer. Indeed, so insatiable is the thirst of capital for profits that it thinks no longer in terms of minutes of workers' time but fractions of minutes. The grotesque precision with which this is done to workers can be seen from a report which appeared in the Daily Mirror of 7 June 1973. It gave particulars of the elements which make up the 32.4 minute rest allowance deal for body press workers on the Allegro car. (See page 16.) The report went on to point out that in a recent dispute, the workers sought to increase the monotony allowance by another 65 seconds! The methods may vary from company to company, or from country to country, but where the profit motive reigns supreme, the degradation and subordination of the worker is the same. George Friedmann has written of two different methods used by great French companies, Berliot in Lyons and Citroen in Paris:

> Why has the Berliot works the reputation, in spite of the spacious beauty of its halls, of being a jail?
>
> Because here they apply a simplified version of the Taylor method of rationalizing labor, in which the time taken by a demonstrator, an "ace" worker, serves as the criterion imposed on the mass of workers. He it is who fixes, watch in hand, the "normal" production expected from a worker. He seems when he is with each worker, to be adding up in an honest way the time needed for the processing of each item. In fact, if the worker's movement seems to him to be not quick or precise enough, he gives a practical demonstration, and his performance determines the norm expected in return for the basic wage.

Add to this supervision in the technical sphere the disciplinary supervision by uniformed warders who patrol the factory all the time and go as far as to push open the doors of the toilets to check that the men squatting there are not smoking, even in workshops where the risk of fire is non-existent.

At Citroen's the methods used are more subtle. The working teams are in rivalry with one another, the lads quarrel over travelling cranes, drills, pneumatic grinders, small tools. But the supervisors in white coats, whose task is to keep up the pace, are insistent, pressing, hearty. You would think that by saving time a worker was doing them a personal favour. But they are there, unremittingly on the back of the foreman, who in turn is on your back; they expect you to show an unheard of quickness in your movements, as in a speeded-up motion picture! Within this context, the desire of companies to recruit only those under the age of 30 can be seen in its dehumanized context.

Although this is the position on the workshop floor, it would be naive indeed to believe that the use of high capital equipment will be any more liberating in the fields of clerical, administrative, technical, scientific and intellectual work.

Some scientists and technologists take the smug view that this can only happen in any case to mere manual workers on the shop floor. They fail to realize that the problem is now at their own doorstep. At a conference on Robot Technology at Nottingham University in April 1973, a programmable drafting or design system was accepted by definition as being a robot. One of the manufacturers of robotic equipment pointed out "Robots represent industry's logical search for an obedient workforce." This is a very dangerous philosophy indeed. The great thing about people is that they are sometimes disobedient. Most human development, technical, cultural and political, depended upon those movements which questioned, challenged and where necessary disobeyed the then established order.

MINIMUM MAINTENANCE FOR THE HUMAN APPENDAGE

The ruling class views all workers whether manual or mental as units of production. Only when that class reality has been firmly grasped can the chasm which divides the potentialities of science and technology from the current reality be understood. The gap between that which is possible and that which is widens daily. Technology can produce a Concorde but not enough simple heaters to save the hundreds of old age pensioners who each winter die in London of hypothermia. Only when one realizes that the system regards old age pensioners as discarded units of production does this make sense—capitalist sense. This is part of their social design, and from a ruling class viewpoint is quite "scientific" and abides closely by the principles observed in machine design.

I know, as a designer, that when you design a unit of production you ensure that you design it to operate in the minimum environment necessary for it to do its job. You seek to ensure that it does not require any special temperature-controlled room unless it is absolutely essential. In designing the lubrication system you do not specify any exotic oils as lubricants unless it is necessary. You ensure that its control system is provided with the minimum brain necessary for it to do its job. You don't, for example, have a machine tape controlled if you can get away with a manual one. Finally, you provide it with the minimum amount of maintenance. In other words, you design for it the maximum life span in which it will operate before a failure. Those who control our society see human beings in the same way. The minimum environment for workers means that you provide them with the absolutely lowest level of housing which will keep them in a healthy enough state to do their job. If one doubts that, it is still worth remembering that 7,000,000 people live in slums in Britain. The equivalent of fuel and lubrication for the machine is the food provided for a worker. This is also kept at a minimum for those who work. We even find Oxford dieticians still telling old age pensioners how they can manage on $4 of food per week.

The minimum brain is provided for the worker by an educational system which imparts enough knowledge to be of use to the industrial system, and which trains him or her to do the job, but does not educate the worker to think about his or her own predicament or that of society as a whole.

The minimum maintenance in Britain is provided through the National Health Service and in the U.S. through Medicaid (if you're poor enough to qualify). Both concentrate on curative rather than preventive medicine. The harsh reality is that when workers have finished their working life they are thrown on the scrap heap like obsolete machines.

If all this is felt to be an extreme position, it is worth recalling the statement of the doctor at Willesden Hospital who said there was no need to resuscitate National Health Patients over the age of 65. (The doctor himself was actually 68.) When a barrage of protest was raised, the statement was hurriedly withdrawn as a mistake! The real mistake he made was to reveal in naked print one of the underlying assumptions of our class divided society. Science and technology cannot be humanely applied in an inherently inhuman society, and the contradictions for scientific workers in the application of their abilities will grow, and if properly articulated will lead to a radicalization of the scientific community.

NEED FOR PUBLIC INVOLVEMENT

Any meaningful analysis of scientific abuse must probe the very nature of the scientific process itself, and the objective role of science within the ideological framework of a given society. As such, it ceases to be merely a "problem of science" and takes on a political dimension. It extends beyond the important, but limited, introverted, soul searching of the scientific community, and recognizes the need for wider public involvement. Many "progressive" scientists now realize that this is so, but still see their role as the interpreters of the mystical world of science for a largely ignorant mass which when enlightened will then support the scientists in their intention "not to use my scientific knowledge or status to promote practices which I consider dangerous."

Those who in addition to being "progressive" have political acumen, know that a "Lysistrata movement," even if it could be organized, is unlikely to terrify monopoly capitalism into applying science in a socially responsible manner. Socially responsible science is only conceivable in a politically responsible society. That must mean changing the one in which we now live.

One of the prerequisites for such political change is the rejection of the present basis of our society by a substantial number of its members, and a conscious political force to articulate that contradiction as part of a critique of society as a whole. The inevitable misuse of science and its consequent impact upon the lives of an ever growing mass of people, provides the fertile ground for such a political development. It should constitute an important weapon in the political software of any conscious revolutionary.

Even Marxist scientists seem to reflect the internal political incestuousness of the scientific community, and demonstrate in practice a reluctance to raise these issues in the mass movement. Thus, the debate has tended to be confined to the rarified atmosphere of the campus, the elitism of the learned body or the relative monastic quiet of the laboratory.

Clearly the ruling class, which has never harbored any illusions about the ideological neutrality of science, will not be over-concerned by this responsible disquiet. The Geneens of I.T.T., and the Weinstocks of General Electric Company do not tremble at the pronouncements of Nobel Laureates. It is true, of course, that the verbal overkill of the ecologist has reverberated through the quality press and caused some concern—not all of it healthy—in liberal circles. But the working class—those who have it within their power to transform society, those for whom such a transformation is an objective necessity—have not as yet been really involved. Yet their day-to-day experience at the point of production brutally demonstrates that a society which strives for profit maximization is incapable of providing a rational social framework for technology (which they see as applied science).

"Socially irresponsible" science not only pollutes our rivers, air and soil, provides C.S. gas for Northern Ireland, produces defoliants for Vietnam and stroboscopic torture devices for police states. It literally also degrades both mentally and physically, those at the point of production, as the objectivization of their labor reduces them to mere machine appendages. The fiancial anaesthetic of the "high wage (a lie in any case) high productivity low cost economy" has demonstrably failed to numb workers' minds to the human costs of the fragmented dehumanized tasks of the production line.

There are growing manifestations in the productive superstructure of the irreconcilable contradictions at the economic base. The sabotage of products on the robot assisted line at General Motors Lordstown plant in the U.S., the 8% absentee rate at Fiat in Italy, the "quality" strike at Chryslers in Britain and the protected workshops in Sweden reveal but the tip of a great international iceberg of seething industrial discontent. That discontent, if properly handled, can be elevated from its essentially defensive, negative stance into a positive political challenge to the system as a whole.

The objective circumstances for such a challenge are developing rapidly as the crushing reality is hammered home by the concrete experience of more and more workers in high capital, technologically based, automated or computerized plants. In consequence, there is a gradual realization by both manual and staff workers that the more elaborate and scientific the equipment they design and build, the more they themselves become subordinated to it, that is to the objects of their own labor. This process can only be understood when seen in the historical and economic context of technological change as a whole.

FUNDAMENTAL DIFFERENCE

The use of fixed capital, that is machinery, and more recently, computers in the productive process marked a fundamental change in the mode of production. It cannot be viewed merely as an increase in the rate at which tools are used to act on raw material. The hand tool was entirely animated by the workers, and the rate at which the commodity was produced—and the quality of it—depended (apart from the raw materials, market forces and supervision) on the strength, tenacity, dexterity and ingenuity of the worker. With fixed capital it is quite the contrary in that the method of work is transformed as regards its use value (material existence) into that form most suitable for fixed capital. The scientific knowledge which predetermines the speeds and feeds of the machine, and the sequential movements of its inanimate parts, the mathematics used in compiling the numerical control program, do not exist in the consciousness of the operator; they are external to him and act upon him through the machine as an alien force. Thus science, as it manifests itself to the workers through fixed capital, although it is merely the accumulation of the knowledge and skill now appropriated, confronts them as an alien and hostile force, and further subordinates them to the machine. The nature of their activity, the movements of their limbs, the rate and sequence of those movements—all these are determined in quite minute detail by the "scientific" requirements of fixed capital. Thus, objectivized labor in the form of fixed capital emerges in the productive process as a dominating force opposed to living labor. We shall see subsequently when we examine concrete situations at the point of production that fixed capital represents not only the appropriation of *living* labor but in its sophisticated forms (computer hardware and software) appropriates the scientific and intellectual output of white collar workers whose own intellects oppose them also as an alien force.

The more, therefore, that workers put into the object of their labor the less there remains of themselves. The welder at General Motors who takes a robotic welding device and guides its probes through the welding procedures of a car body is on the one hand building skill into the machine, and de-skilling himself on the other. The accumulation of years of welding experience is absorbed by the robot's self programming systems and will never be forgotten. Similarly, a mathematician working as a stressperson in an aircraft company may design a software package for the stress analysis of airframe structures and suffer the same consequences in his or her job. In each case they have given part of themselves to the machine and in doing so have conferred life on the object of their labor—but now this life no longer belongs to them but to the owner of the object.

Since the product of their labor does not belong to the workers, but to the owner of the means of production in whose service the work is done and for whose benefit the product of labor is produced, it necessarily follows that the object of the workers' labor confronts them as an alien and hostile force, since it is used in the interests of the owner of the means of production. Thus, this "loss of self" of the worker is but a manifestation of the fundamental contradictions at the economic base of our society. It is a reflection of the antagonistic contradiction between the interest of capital and labor, between the exploiter and the exploited. Fixed capital, therefore, at this historical stage is the embodiment of a contradiction, namely that the means which could make possible the liberation of the workers from routine, soul-destroying, backbreaking tasks, is simultaneously the means of their own enslavement.

IS "POLITICAL" CHANGE ENOUGH?

It is therefore, obvious that the major contradiction can only be resolved when a change in the ownership of the means of production takes place. Much less obvious, however, is whether there exists a contradiction (non-antagonistic) between science and technology in their present form and the very essence of humanity. It is quite conceivable that our scientific methodology, and in particular our design methodology, has been distorted by the social forces that give rise to its development. The question, therefore, must arise whether the problems of scientific development and technological change, which are *primarily* due to the nature of our class-divided society, can be solved solely by changing the economic base of that society.

The question is not one of mere theoretical and academic interest. It must be a burning issue in the minds of those attempting to build a people's democracy. It must be of political concern to them to establish whether Western technology can be simply applied to a socialist society. Technology at this

Implications of New Technology 75

historical stage, in this class-divided society, is the embodiment of two opposites—the possibility of freeing workers yet the actuality of ensnaring them. The possibility can only become actuality when the workers own the objects of their labor. Because the nature of this contradiction has not been understood, there have been the traditional polarized views "technology is good" and "technology is bad". These polarized views are of long standing and not merely products of space-age technology. From the earliest times a view has persisted that the introduction of mechanization and automated processes would automatically free people to engage in creative work. This view has persisted as consistently in the field of intellectual work as it has in that of manual labor. As far back as 1624, when Pascal introduced his first mechnical calculating device he said, "I submit to the public a small machine of my own invention, by means of which you alone may without any effort perform all the operations of arithmetic and may be relieved of the work which has so often fatigued your spirit when you have worked with the counters and with the pen." Only 28 years earlier in 1596 an opposite view was dramatically demonstrated when the city council of Danzig hired an assassin to strangle the inventor of a labor-saving ribbon loom. This defensive, if understandable, attempt was repeated time and again in various guises during the ensuing five hundred years to resolve a contradiction at an industrial level when only a revolutionary political one would suffice. It is, of course, true that the contradiction manifests itself in industrial forms even to this day.

THE DEDICATED APPENDAGE

It has been common for some time to talk about "dedicated machines". It is now a fact that when defining a job function employers define a dedicated appendage to the machine, the operator.

Even our educational system is being distorted to produce these "dedicated men for dedicated machines." It is no longer a matter that the people are being educated to think, they are being trained to do a narrow, specific job. Much of the unrest among students is recognition that they are being trained as industrial fodder for the large monopolies in order to fit them into narrow fragmented functions where they will be unable to see in an overall panoramic fashion the work on which they are engaged.

In order to ensure that the right kind of "dedicated product" is turned out of the university, we find the monopolies attempting to determine the nature of university curricula and research programs. Warwick University was a classical example. In particular, at research level, the monopolies increasingly attempt to determine the nature of research through grants which they provide to universities while research scientists still harbor illusions that they are in practice "independent, dedicated searchers after truth."

The "truth" for them has to coincide with the interests of the monopolies if they are to retain their jobs. William H. Whyte, Jr. in the "Organization Man" pointed out that in the United States, out of 600,000 persons then engaged in scientific research, not more than 5,000 were allowed to choose their research subject and less than 4 per cent of the total expenditure was devoted to "creative research" which does not offer immediate prospects of profits.

He recognized the long term consequences of this and concluded,

> If corporations continue to mold scientists the way they are now doing, it is entirely possible that in the long run this huge apparatus may actually slow down the rate of basic discovery it feeds on.

PROBLEMS FOR THE EMPLOYER

I have up to now concentrated on the contradictions as they affect the worker by hand or brain. There are, of course, problems for the employer and an understanding of some of these is of considerable tactical importance.

One of the contradictions for the employer is that the more capital he accumulates in any one place the more vulnerable it becomes. The more closely he synchronizes his industry and production by using computers the greater becomes the strike power of those employed in it. Mao Tse Tung once said, in his military writings, that the more capitalized an army becomes the more vulnerable it becomes also. This has been demonstrated in Vietnam, where an N.L.F. cadre with a $3.00 shell could destroy an American aircraft with an airborne computer costing something like $5 million.

A Palestine guerrilla with a revolver costing perhaps $40 can hijack a plane costing several million dollars and destroy it at some safe airfield. High capital equipment, although it appears all powerful and invincible, always has a point of vulnerability and possibilities for sabotage and guerrilla warfare are considerable. A quite small force can destroy or immobilize plant equipment or weapons costing literally millions. The capitalization of industry also produces an analogous situation. In the past, when a clerical worker went on strike it had precious little effect. Now if the wages of a factory are calculated by a computer a strike by clerical workers can disrupt the whole plant. It is also true on the factory floor that in the highly synchronized motor car industry a strike of 12 workers in the foundry can stop large sections of the entire motor car industry.

The same is happening in the design area. As high capital equipment, through Computer Aided Design, is being made available to design staffs, firstly it proletarianizes them, but secondly it also increases their strike power. In the past when a draftsman went on strike he simply put down his 6H pencil and his eraser, and there was unfortunately a considerable length of time before an effect was felt upon production, even when the manual workers were blacking his drawings. With the new kind of equipment described, where numerical controlled tapes are being prepared or where high capital equipment is used for interactive work, the effects of a strike will in many instances be immediate, and production will be affected in a very short length of time.

PARITY WITH THE MACHINE

This will apply equally to hosts of other jobs and occupations in banking, insurance, power generation, civil transport, as well as those more closely connected to industry and production.

Thus, while the introduction of fixed capital enables the employer to displace some workers and subordinate others to the machine it also embodies within it an opposite in that it provides the worker with a powerful industrial weapon to use against the employer who introduced it.

This is even the case when industrial action short of strike action is taken. As has been pointed out the activity of the worker is transformed to suit the requirement of fixed capital. The more complete that transformation, the greater is the disruptive effect of the slightest deviation by the worker from his pre-determined work sequence. Industrial militants with an imaginative and creative view of industrial harassment have been able to exploit this contradiction by developing such techniques as "working to rule," "working without enthusiasm" and "days of non-co-operation." I know from personal experience that these techniques can reduce the output of both manual and staff workers by up to 70 per cent without placing on the workers involved the economic hardship of a full strike.

Since much of the sophisticated equipment I have described earlier is very sensitive and delicate in a scientific sense it has to be handled with great care and is accommodated in purpose-built structures in conditions of clinical cleanliness. In many industries the care the employer will lavish on his fixed capital is in glaring contrast with the comparatively primitive conditions of his living capital. The campaign for parity with equipment which perhaps started facetiously in 1964 with that placard at Berkeley which parodied the IBM punchcard ("I am a human being: Please do not fold, spindle or mutilate") has now assumed significant industrial dimensions. In June 1973 designers and draftsmen members of the AUEW-TASS employed by a large Birmingham engineering firm, officially claimed "Parity of environment with the C.A.D. equipment" in the following terms:

> This claim is made in furtherance of a long standing complaint concerning the heating and ventilation in the Design and Drawing Office Area going back to April 1972. Indeed to our certain knowledge these

working conditions have been unsatisfactory as far back as 1958. We believe that if electromechanical equipment can be considered to the point of giving it an air conditioned environment for its efficient working, the human beings who may be interfaced with this equipment should receive the same consideration.

It is an interesting reflection on the values of advanced technological society that it subsequently took three industrial stoppages to achieve for the designers conditions approaching those of the C.A.D. equipment. The exercise also helped to dispel some illusions about highly qualified design staff not needing trade unions.

Scientists must now begin to learn the lessons of such experiences, and to understand that their destiny is bound up with all of those "molded" by the systems. Only when they are prepared to be involved in political struggle with them, can they ever begin to move towards a society where the scientist will be able to give "according to his ability". It is the historical task of the working class to effect such a transformation, but in that process scientists and technologists can be powerful and vital allies for the working class as a whole. This will mean that scientists will have to involve themselves in the political movement. Above all, they must attempt to understand that the products of their ingenuity and scientific ability will become the objects of their own oppression and that of the mass of the people until they are courageous enough to help form that sort of society "when the enslaving subordination of the individual to the division of labour, and with it the antithesis between mental and physical work has vanished, when labour is no longer merely a means of life but has become life's principal need, when the productive forces have also increased with the all round development of the individual, and all the springs of cooperative wealth flow more abundantly. Only then will it be possible completely to transcend the narrow outlook of the bourgeois right, and only then will society be able to inscribe on its banners—"From each according to his ability, to each according to his needs." Then, and then only will scientists be able to truly give of their ability to meet the needs of the community as a whole rather than maximize profits for the few.[15]

FOUR

Drawing Up the Corporate Plan at Lucas Aerospace

At no stage in human history has the potential for solving our economic problems been so great. Human ingenuity, expressed through appropriate science and technology, could do much to free our world from squalor and disease and fulfil our basic needs of food, warmth and shelter. Yet at the same time there is a growing disquiet, even alarm, among wide sectors of the community about the future of "industrial society."

THE CONTRADICTIONS

There are many contradictions which highlight the problems of our so-called technologically advanced society. Four of these contradictions in particular influenced the events at Lucas Aerospace.

First, there is the appalling gap which now exists between that which technology could provide for society, and that which it actually does provide. We have a level of technological sophistication such that we can design and produce the Concorde, yet in the same society we cannot provide enough simple heating systems to protect old age pensioners from hypothermia. In the winter of 1975-6, 980 people died of the cold in the London area alone.

We have senior automotive engineers who sit in front of computerized visual display units "working interactively to optimise the configuration" of car bodies to make them aerodynamically stable at 120 miles an hour when the average speed of traffic through New York is 6.2 miles an hour. It was in fact 11 m.p.h. at the turn of the century when the vehicles were horsedrawn. In London at certain times of the day it is about 8.5 miles an hour.

We have sophisticated communications systems such that we can send messages round the world in fractions of a second, yet it now takes longer to send a letter from Washington to New York than it did in the days of the stage coach. Hence we find on the one hand the linear drive forward of complex esoteric technology in the interests of the multinational corporations, and on the other hand, the growing deprivation of communitites and the mass of people as a whole.

The second contradiction is the tragic wastage our society makes of its most precious asset—that is the skill, ingenuity, energy, creativity and enthusiasm of its ordinary people. We now have in Britain 1.6 million people out of work. There are thousands of engineers suffering the degradation of unemployment when we urgently need cheap, effective and safe transport systems for our cities. There are thousands of electricians robbed by society of the right to work when we urgently need economic urban heating systems. We have, I believe, 180,000 building workers out of a job when by the government's own statistics it is admitted that about 7 million people live in semi-slums in this country. In the London area we have about 20 percent of the schools without an operating indoor toilet when the people who could be making these are rotting away on the unemployment line.

The third contradiction is the myth that computerization, automation and the use of robotic devices will automatically free human beings from soul destroying backbreaking tasks and leave them free to engage in more creative work. The perception of my trade union colleagues and that of millions of workers in the industrial nations is that in most instances the reverse is actually the case.

At an individual level, the totality that is a human being is ruthlessly torn apart and its component parts set one against the other. The individual as producer is required to perform grotesque alienated tasks to produce throwaway products to exploit the individual as consumer. We are at a stage where our incorporate science and technology, with its concepts of efficiency and optimization, converge with the requirements and value systems of the vast multinational corporations.

Fourth, there is the growing hostility of society at large to science and technology as at present practiced. If you go to gatherings where there are artists, journalists and writers and you admit to being a technologist, they treat you as some latter day Yahoo—to misquote Swift. They really seem to believe that you specified that rust should be sprayed on car bodies before the paint is applied; that all commodities should be enclosed in non recyclable containers; and that every large scale plant you design is produced specifically to pollute the air and the rivers. There seems to be no understanding of the manner in which scientists and technologists are used as mere messengers of the multinational corporations, whose sole concern is the maximization of profits. It is therefore not surprising that some of our most able and sensitive high school students will not now study science and technology because they perceive it to be such a dehumanized activity in our society.

Propelled by the frantic linear drive forward of this form of science and technology, we witness the exponential change in the organic composition of capital and the resultant growth of massive structural unemployment. So stark is the situation becoming that predictions of 20 million jobless in the E.E.C. countries by 1990 no longer seem absurd.

LUCAS WORKERS RESPOND

All these four contradictions, and indeed many others, have impacted themselves upon us in Lucas Aerospace over the past five years. We do work on equipment for the Concorde, we have experienced structural unemployment, and we know day by day of the growing hostility of the public to science and technology.

Lucas Aerospace was formed in the late 1960s when parts of Lucas Industries took over sections of GEC, AEI and a number of other small companies. It was clear that the company would engage in a rationalization program along the lines already established by Arnold Weinstock in GEC. This, it will be recalled, was the time of Harold Wilson's "white heat of technological change." The tax-payer's money was being used through the Industrial Reorganisation Corporation to facilitate this rationalization program. No account at all was taken of the social cost, and Arnold Weinstock subsequently sacked 60,000 highly skilled workers.

We in Lucas Aerospace were fortunate in the sense that this happened about one year before our company embarked on its rationalization program. We were therefore able to build up a Combine Committee which would prevent the company setting one site against the other in the manner Weinstock had done. This body, the Combine Committee, is unique in the British Trade Union Movement in that it links together the highest level technologists and the semi-skilled workers on the shop floor. There is therefore a creative cross fertilization between the analytical power of the scientist and perhaps what is more important, the direct class sense and understanding of those on the shop floor.

As structural unemployment began to affect us, we looked around at the manner in which other groups of workers were attempting to resist it. In Lucas we had already been engaged in sit-ins, in preventing the transfer of work from one site to another and a host of other industrial tactics which had been developed over the past five years, but we realized that the morale of a workforce very quickly declines if the workers can see that society, for whatever reason, does not want the products that they make. We therefore evolved the idea of a campaign for the right to work on socially useful products.

It seemed absurd to us that we had all this skill and knowledge and facilities at the same time that society urgently needed equipment and services which we could provide, and yet the market economy seemed incapable of linking the two. What happened next provides an important lesson for those who wish to analyze how society can be changed.

AN IMPORTANT LESSON

We prepared 180 letters which described in great detail the nature of the workforce, its age, its skills, its training and the machine tools, equipment and laboratories that were available to us together with the types of scientific staff and the design capabilities which they possessed. The letters went to 180 leading authorities, institutions, universities, trade unions and other organizations, all of which had in the past, one way or another, suggested that there was a need for the humanization of technology and the use of technology in a socially responsible way. What happened was really a revelation to us in Lucas. All of these people who had made great speeches up and down the country, and in some cases written voluminous books about these matters, were smitten into silence by the specificity of our request. We had asked them quite simply, "What could a workforce with these facilities be making that would be in the interests of the community at large?" and they were silent— with the exception of four individuals, Dr. David Elliot at the Open University, Professor Thring at Queen Mary College, Richard Fletcher and Clive Latimer, both at the North East London Polytechnic.

We then did what we should have done in the first place. We asked our own members what they thought they should be making.

I have never doubted the ability of ordinary people to cope with these problems, but not doubting is one thing, having concrete evidence is something quite different. That concrete evidence began to pour in within three or four weeks. In a short time we had 150 ideas of products which we could make and build with the existing machine tools and skills we had in Lucas Aerospace. We elicited this information through our shop stewards' committees via a questionnaire.

I should explain that this questionnaire was very different from those which the soap powder companies produce, where the respondent is treated as some kind of passive cretin. In our case, the questionnaire was dialectically designed. By that I mean, that in filling it in, the respondent was caused to think about his or her skill and ability, the environment in which he or she worked and the facilities available to them. We also deliberately composed it so that they would think of themselves in their dual role in society, that is, both as consumers and as producers. We were, therefore, deliberately transcending the absurd division which our society imposes upon us, which seems to suggest that there are two nations, one that works in factories and offices, and an entirely different nation that lives in houses and communities. We pointed out that what we do during the day at work should also be meaningful in relation to the communities in which we live. We also deliberately designed the questionnaire to cause the respondents to think of products for their use value and not merely for their exchange value.

When we collected all these proposals we refined them into six major product ranges which are now embodied in six volumes each of approximately 200 pages. They contain specific technical details, economic calculations and even engineering drawings. We quite deliberately sought a mix of products which included, on the one hand, those which could be designed and built in the very short term and, on the other, those which would require long term development; those which could be used in metropolitan Britain mixed with those which would be suitable for use in the Third World. Products which incidently could be sold in a mutually non-exploitative fashion. Finally, we sought a mix of products which would be profitable by the present criteria of the market economy and those which would not necessarily be profitable but would be highly socially useful.

THE PRODUCTS AND IDEAS

Before we even started the Corporate Plan, our members at the Wolverhampton Plant visited a center for children with spina bifida and were horrified to see that the only way the children could propel themselves about was literally by crawling on the floor, so they designed a vehicle which subsequently became known as HOBCART. It was highly successful, and the Spina Bifida Assocation of Australia wanted to order 2,000 of these. Lucas would not agree to manufacture them because, they said, it was incompatable with their product range. At that time the Corporate Plan was not developed and so we were unable to press for the manufacture of HOBCART. However, the design and development of this product were significant in another sense. Mike Parry Evans, its designer, said that it was one of the most enriching experiences of his life when he carried the hobcart down and saw the pleasure on the child's face. It meant more to him, he said, than all the design activity he had been involved in up till then. For the first time in his career he actually saw the person who was going to use the product he had designed. It was enriching also in another sense, because he was intimately in contact with a social human problem. He needed to make a clay mold of the child's back so that the seat would support it properly. It was fulfilling in that he was working in a multi-disciplinary team together with a medical type doctor, a physiotherapist and a health visitor. I mention this because it illustrates very graphically that it is not true to suggest that aerospace technologists are only interested in complex esoteric technical problems. It can be far more rewarding for them if they are allowed to relate their technology to human and social problems.

A LIFE SUPPORT SYSTEM

Some of our members at another plant realized that about 30 percent of the people who die of heart attacks die between the point at which the attack occurs and the stage at which they are located in the intensive care unit in the hospital. They designed a light, simple, portable life support system which can be taken in an ambulance or at the side of a stretcher to keep the patient "ticking ove " until they can be linked to the main life support system in the hospital.

They also learned that many patients died under critical operations because of the problems of maintaining the blood at a constant optimum temperature and flow. This, it seemed to them, was a simple technical problem once one was able to get behind the feudal mysticism of the medical profession. They designed for this a simple heat exchanger and pumping system and they built it in prototype. I understand that when the assistant chief designer at one of our plants had to have a critical operation, they were able to convince the local hospital to use it, and it was highly successful.

ENERGY CONSERVING PRODUCTS

In the field of alternative energy sources we have come up with a very imaginative range of proposals. It seemed to us absurd that it takes more energy to keep New York cool during the summer than it does to heat it during the winter. If, therefore, there were systems which could store this energy when it was not required and use it at a time when it was required, this would make a lot of sense.

One of the proposals for storing energy in this way was to produce gaseous hydrogen fuel cells. These would require considerable funding from the government, but would produce a means of conserving energy which would be ecologically desirable and socially responsible.

There are further designs for a range of solar collecting equipment which could be used in low energy houses. We worked on this in conjunction with Clive Latimer and his colleagues at the North East London Polytechnic, and components for a low energy house were produced. I should add that this house was specifically designed so that it could be constructed on a self-build basis. In fact, some of the students working on the Communications Design Degree Course at that polytechnic are now writing an instruction manual which would enable people without any particular skills to go through a learning process, and at the same time produce very ecologically desirable forms of housing. One can now see that if this concept were linked to imaginative government community funding, it would be possible, in areas of high unemployment where there are acute housing problems, to provide funds for employing those in that area to build houses for themselves.

We have made a number of contacts with county councils, as we are very keen to see that these products are used in communities by ordinary people. We are unhappy about the present tendency in alternative technology for products to be provided which are little more than playthings for the middle class in their architect-built houses. Hence we have already made links via the Open University with the Milton Keynes Corporation, and have designed, and are currently building in conjunction with the O.U., some prototype heat pumps which will use natural gas and will increase the actual coefficient of performance (C.O.P.) to 2.8 at 0^0C and 4.0 at 10^0

A NEW HYBRID POWER PACK

The problem of finding an ecologically desirable power unit for cars is one which needs to be solved as a matter of urgency.

Lucas Electrical, which is a separate company from Lucas Aerospace, has proposed a solution based on a battery driven car. However, with a vehicle of this kind it is necessary to recharge it approximately every 40 miles of stop-start driving and every 100 miles on flat terrain. Furthermore, it is necessary to carry a significant weight of batteries. At a chassis weight of around 1300 kilos an additional 1000 kilos of batteries would have to be carried. Because the batteries need charging at regular intervals, vehicles of this kind are unsuitable for random journeys. Additionally, there would be a need for a new roadside infrastructure with frequent charging facilities.

One possibility would be to provide these in existing garages, but having a large number of vehicles waiting to be charged overnight creates considerable difficulties. This gives rise to the notion that interchangeable batteries should be available. However, it would clearly be a significant task to regularly change 1000 kilos of batteries (about a ton). Moreover, providing storage space for large numbers of batteries in the

London area is reckoned to cost around $80 to $120 per cubic foot per annum. Thus drivers would have to pay for the additional reserve batteries *and* for the space to store them.

The aerospace workers' approach was quite different. They pointed out that the average vehicle has an engine twice and maybe three times larger than is necessary simply to give it take-off torque. Once the vehicle is moving along, a much smaller engine could satisfactorily power it. They also pointed out that the performance characteristics of an electric motor are the dialectical opposite of those of a petrol engine. That is to say, the electric motor has a high starting torque whereas a petrol engine has a better torque at high revs. By linking these two together a new unity can be formed. A small component combustion engine, running constantly at its optimum revs. and at its optimum temperature, drives a generator which in turn charges a very small stack of batteries. These act merely as a temporary energy store and supply power to an electric motor which drives the transmission system, or in a revised version, will drive hub motors directly on wheels.

A number of variations on this theme have been proposed; one is for inter-city driving. Once the vehicle has gained speed, the combustion engine could drive the wheels directly through the mechanical transmission system, whereas when the vehicle enters the suburban area, with the consequent stop-start driving, it could run in the hybrid mode.

The Lucas workers also envisaged that in coming years the internal combustion engine would be banned from city centers. With the hybrid power pack, it would be possible to drive to the perimeter of the prohibited zone and then within the zone drive slowly, solely on stored energy. Subsequently, the system would be recharged when operating in the hybrid mode elsewhere.

In general use, however, the internal combustion engine would be running continuously at its constant optimum rate. All the energy that is wasted as one starts from cold, accelerates and decelerates, changes gear or idles at traffic lights, would go into the system as useful energy. This, it is suggested, would improve fuel consumption by about 50%. Since the engine is running constantly and at its optimum temperature, it follows that combustion of the gases will be much more complete, thereby reducing the emission of toxic fumes by about 80% since the unburned gases are not going out into the atmosphere, and it would improve specific fuel consumption by 50%. The initial calculations on this have subsequently been supported by work done in Germany. The engine would run at constant revs so the resonance frequencies of the various components in the system could be different from that of the engine and noise levels reduced. With a background noise of 65 decibels, the power pack would be inaudible 10 metres away. A prototype unit of this kind is now being built and tested under the direction of Professor Thring at Queen Mary College, London.

No individual component of the system is itself revolutionary. What is new is the creative manner in which the various elements have been put together. It may be asked, of course, why such a power pack had not been designed and developed before. The simple answer it seems to us, is that such a power pack would have to last for 15 years or so, and maintenance services would have to be developed to repair and maintain them. This is completely contrary to the whole ethos of automotive design which has at its basis the notion of non-repairable throwaway product with all the terrible waste of energy and materials which that implies.

While the Lucas Workers are proposing this kind of power pack, their colleagues at another large car manufacturer are having to design and develop an engine which would be thrown away after 20,000 miles or two years, whichever comes first. The idea is that the engine would simply be bolted on the input side of the gearbox so that you would simply connect the carburetor to it. The owner would even be denied the pleasure of putting water and oil in! It is a criminally irresponsible type of technology, yet the whole political and economic infrastructure of society is based on the assumptions of this technology—namely, that the rate of obsolescence of products will increase, and that the rate of production and consumption will grow. We are convinced that Western society cannot carry on in this wasteful and arrogant way much longer.

ALL PURPOSE POWER GENERATION

Drawing on our aerodynamics knowhow, we have proposed a range of wind generators. In some instances, these would have a unique rotor control in which the liquid used as the medium for transmitting the heat is also used to effect the braking, and is thereby heated in the process itself.

We have proposed a range of products which would be useful in third world countries. We feel, incidentally, that we should be very humble about suggesting that our kind of technology would be appropriate in these countries. If one looks at the incredible mess we have made of technology in our society, probably one of the most important things the Third World countries could learn from us is what not to do rather than what to do! It is also a very arrogant assumption to believe that the only form of technology is that which we have in the West. I can see no reason why there should not be technologies which are compatible with the cultural and social structures of these other countries.

At the moment, our trade with these countries is essentially neocolonialist. We seek to introduce forms of technology which will make them dependent upon us. When the gin and tonic brigade go out to sell a power pack, for example, they always seek to sell a dedicated power pack for each application; that is, one power pack for generating electricity, another power pack for pumping water and so on.

The Lucas workers' approach is quite different. They have designed a universal power pack which is quite capable of providing a wide range of services. It has a basic prime mover which could run on a wide range of fuels, including naturally available materials, methane gas and so on.

By using a specially designed, variable speed gearbox it is possible to vary the output speed over a very wide range. Thus the unit is capable of providing the speed and power necessary to drive a generator which could supply electricity at night. When running at one of the lower speeds, it could drive a compressor to provide compressed air for pneumatic tools. It could also drive a hydraulic pump to provide power for lifting equipment, while at very low speeds it could drive a water pump and be used for irrigation purposes. The unit could thus be used in a number of ways for almost 24 hours a day.

In considering the design, the various bearing surfaces have been made much larger than normal and the components deliberately designed in such a way as to make them last for about 20 years with almost no maintenance. It has also been designed in such a way that if maintenance is necessary, it can be carried out using local skills. Furthermore, instructions would be such that these maintenance processes would help to develop self reliance among those using the equipment.

ROAD/RAIL VEHICLE

In the mid 50s Lucas Aerospace (Rotax) spent over one million pounds developing an actuating mechanism whereby a set of pneumatic tires could be brought down into position so that a railway coach could run on the roads. In its railway mode, a metal rim would run on a metal track, which in practice resulted in all the shocks going up through the superstructure. Inevitably, this meant a large rigid superstructure of the type we have inherited from Victorian rolling stock design.

But again, there is another approach which was followed up by the Lucas workers and Richard Fletcher and his colleagues at the North East London Polytechnic. By using a small guidewheel running on the rail, with servo mechanism feedback to the running wheels, the wheel can be steered along the track while the pneumatic tires run on the rails.

With the guide mechanisms retracted, of course, the vehicle can be used conventionally on the road. This provides the basis of a flexible light weight vehicle which is capable of going up a rail incline of one in six.

Normal railway stock, because of the low friction between the metal rim and the metal track, is capable of going up an incline of no more than one on 80. This means that when a new railway line is being laid, for example, in the developing countries, it is necessary literally to flatten the mountains and fill up the valleys, or as an alternative, build tunnels and viaducts. Typically, this costs over $2 million per track mile. With the hybrid vehicle, running on rubber wheels and having the capacity to climb an incline of one in 6, it is possible to follow the natural terrain and lay down new railway lines for $44,000 per track mile. The vehicle can, of course, be run on disused tracks to service remote areas.

A prototype of the road-rail vehicle has been built at the North East London Polytechnic and tested out on the East Kent railway line with great success. In parts of Britain, there is a growing interest in a vehicle of this kind since it could provide the basis for a truly integrated transport system with vehicles running through our cities like coaches and then moving straight on to the railway network.[2]

KIDNEY MACHINES

The Lucas workers do not merely design and build *new* products. There are one or two existing products in Lucas Aerospace which they would like to see produced at a much greater rate. One of those is the home dialysis or kidney machine. About four years ago, the company attempted to sell off its kidney machine division to an international company operating from Switzerland. We were able to prevent them doing so at that time both by threats of action and the involvement of some M.P.s. When we checked on the requirements for kidney machines in Britain we were horrified to learn that 3,000 people die each year because they cannot get a machine. In the Birmingham area, if you are under 15 or over 45 you are allowed, as a medical practitioner put it so nicely, "to go into decline." The doctors sit like judge and juries with the governors of hospitals deciding who will be saved. One doctor told us how distressed he was by this situation and admitted that sometimes he did not tell the families of the patients that this was happening as it would be too upsetting for them.

We were disgusted when we saw, in a recent I.T.V. program, an interview with a teacher who was over 45 and being allowed to go into decline. She said she was going to commit suicide at some stage so that her grandchildren would not see her going through the progressive stages of debility. Ernie Scarbrow, the Secretary of the Combine Committee said:

It is outrageous that our members in Lucas Aerospace are being made redundant when the state has to find them £40 a week to do nothing except suffer the degradation of the dole queue. In fact the £40 a week amounts to about £70 a week when the cost of administration is taken into account. Our workers should be given this money and allowed to produce socially useful products such as the kidney machine. Indeed, if the social contract had any meaning and if there were such a thing as a social wage surely this is the kind of thing which it should imply, namely having foregone wage increases in order that we could expand medical services, we should then have the opportunity of producing medical equipment the community requires.

TELECHIRIC DEVICES

One of the most important political and technological proposals in the Corporate Plan is for the design of "telechiric" (hands at a distance) devices. With these systems, the human being would be in control, realtime, all the time, and the system would merely mimic human activity, but not objectivize it. Thus the producer would dominate production, and the skill and ingenuity of the worker by hand and brain would be central to the activity and would continue to grow and develop. This would link human intelligence with advanced technology, and help to reverse the historical tendency to objectivize human knowledge and thereby confront the worker with an alien and opposite force as described earlier.

The methods by which the Lucas workers arrived at the concept of this product range are in themselves revealing. It was suggested to them that it would be highly socially responsible if a means could be found of protecting maintenance workers on North Sea Oil Pipelines. These maintenance workers experience a very high accident rate because of the depths at which they have to work.

Since they had been conditioned by traditional design methods, the Lucas workers immediately thought of a robotic device which would eliminate the human being completely.

However, the more they began to consider the programming problems of getting a system which would recognize which way a hexagon nut was about (much less if it had a barnacle on it!) and select the correct spanner and apply the correct torque, they began to recognize what a difficult task this would be. Yet it is the kind of task which a skilled worker can perform even "without thinking about it." They simply have to look at the diameter of a nut and bolt and will know through years of experience what torque they can apply to it without wringing it off, and yet, at the same time tighten it sufficiently so that it won't become loose again. This they can do without any "scientific knowledge" such as the torsional rigidity of the bolt or the shear strength of its material, yet they will get it right repeatedly. As Polanyi once put it, "There are things we know but cannot tell." That is to say, workers don't express this knowledge in its written and spoken form. They demonstrate their knowledge and intelligence by what they do!

Indeed, the Lucas workers began to realize that in Western society we are confusing linguistic ability with intelligence. They then began to compare the levels of intelligence of robotic equipment of that kind with total human information processing capability, and while comparisons of that kind are notoriously difficult, the order of things is the machine 10^3 to 10^4 and the human being 10^{14}. This 10^{14} brings with it, however, consciousness, will, imagination, ideology, political aspirations; and these are precisely the attributes which employers regard as disruptive.

PEOPLE ARE TROUBLE—MACHINES OBEY

One doesn't have to be a sociological Einstein to work this out. The multinationals and the employers are so arrogant that they actually put it in writing in case we might fail to understand. Thus the *Engineer* had a headline recently which said, "People are trouble, but machines obey."[2] It is, therefore, no accident that systems are designed around the trivial 10^3 whereas the 10^{14} is deliberately suppressed. It is a political act which reflects the power relationships at the point of production.

The Lucas workers feel that there are hazardous, dangerous jobs which should be automated out of existence. What they are questioning is the politics of elevating these design methods to universal principles.

It has been pointed out that technological change viewed thus has more to do with the exercise of control over the workforce than it has to do with increasing productivity.[3] Indeed, Andrew Ure, in his "Philosophy of Manufactures" put it even more clearly "(the industrialists aim to stop any) process which requires peculiar dexterity and steadiness of hand, from the cunning workman, and put it in charge of a mechanism so self regulating, that a child may superintend it. The grand object therefore, of the manufacturer is, through the union of capital and science, to reduce the task of his workpeople to the exercise of vigilance and dexterity (appropriate to a child)."

The extent to which capital and science has succceeded in achieving this was dramatically illustrated in the July '79 issue of the *American Machinist*. It reported that an engineering firm had found that the ideal operators for its numerically controlled machining center were mentally handicapped workers.

One of the workers held up as ideal for this type of work had a maximum intelligence level of a 12 year old. The employer pointed out in gloating terms, "He loads every table exactly the way he has been taught, watches the Moog operate and then unloads. It's the kind of tedious work that some non-handicapped individual might have difficulty in coping with."

Clearly, it might have been highly laudible had the ojective been to provide work for the mentally handicapped, but what happens in installations of this kind is that some of the most highly skilled, satisfying and creative work on the shop floor, such as turning and milling, is so deskilled by these new technologies that it is rendered suitable for 12 year olds.

This historical process of deskilling[4] is an important means through which the employer extends his control over his employees. But in a wider sense, it destroys the social and cultural values which surround the exercise of those skills and the means by which they are acquired. Indeed, we seem to have seriously underestimated the educational, cultural and other significance of skill and craftsmanship.[5]

Thus, the significance of raising these issues through the very specific proposals surrounding the telechiric devices. The Lucas workers are indeed developing profound political ideas; so also are those who, in the wider sense, are proposing human centered systems even in the field of high level intellectual activity such as design.

HUMAN CENTERED SYSTEMS

Howard Rosenbrock, a highly creative scientist, has developed advanced interactive graphic systems which actually place the designer and human intelligence at the center of the design process. It is he who makes the observation, "My own conclusion is that engineering is an art rather than a science; and by saying that I imply a higher, not a lower status."

In the field of Computer Aided Design, he cautions against the computer becoming an automated design manual, leaving only minor choices to the design engineer. The automated design manual approach, he suggests, "...seems to me to represent a loss of nerve, a loss of belief in human abilities and further unthinking application of the doctrine of the Division of Labour."

The designer is reduced to making a series of routine choices between fixed alternatives, in which case "His skill as a designer is not used, and decays."[6]

Rosenbrock has developed graphic displays from which the designer can assess stability, speed of response, sensitivity to disturbance and other properties of the system.[7] This he and his colleagues did by using the inverse Nyquist Array. Having demonstrated through his computer aided design system that different value systems can be applied to problems of this kind, Rosenbrock then raises the much wider question as to whether we are not cutting off options in other fields of intellectual work in rather the same way that we have done at an earlier historical stage in the field of manual work. He has termed this the "Lushai Hills Effect."

Other computer scientists, such as Weizenbaum, are now seriously questioning where their work is taking society and what its impact is upon human beings and their self image.[8]

What is often lacking in honest expressions of concern of this kind is an economic and political analysis of the forces in society which control and distort science and technology to fulfill specific class roles.

Thus, the discussions at Lucas and elsewhere should really be viewed in a much wider context of the overall questioning of the way science and technology is developing under advanced capitalist society. It is linked with the wider challenges workers are attempting to make against the way technology is being organized: the Green Bans Movement* in Australia, the attempts by Fiat workers in Italy to transcend the narrow economism which characterized trade union activity for so long, and the courageous struggle by the women at Algots Nord, a clothing factory in Northern Sweden.[9]

All of these forces can be linked together in a challenge to the system as a whole and can act as a forerunner in a transformation of society which will take it away from its present exploitative, hierarchal form to a new type of society. About this new society the founder of cybernetics, Norbert Weiner, once said:

> [It] differs from those propounded by many fascist successful businessmen and politicians. People of this type prefer an organisation where all information emanates from the top and where there is no feedback. The subordinates are degraded to become effectors of an alleged higher organism. It is easier to set in motion a galley or factory in which human beings are used to a minor part of their full capacity only, rather

---

*The Green Ban is a refusal by building laborers to demolish buildings of historical or community value, or to build on areas of environmental concern, or other industrial action to prevent the despoiling of cities or the countryside. There was even a green ban on a beautiful old fig tree in the main park in Sydney which the planners wanted to remove so that they could put up a block of offices.

than create a world in which these human beings may fully develop. Those striving for power believe that a mechanised concept of human beings constitutes a simple way of realising their aspirations to power. It maintains that this easy way to power, not only destroys all ethical values in human beings, but also our very slight aspirations for the continued existence of mankind.

The new technologies highlight the fact that we are at a unique historical turning point. We must not allow our common sense to be bludgeoned into silence by technocratic and scientific jargon, nor should we be intimidated by the determinism of science and technology into believing that the future is already fixed. The future is not "out there" in the sense that America was out there before Columbus went to discover it. It has yet got to be built by human beings, and we do have real choices, but these choices will have to be fought for, and the issues are both technical and political. If we ignore this we may find "All our inventions and progress seem to result in endowing material forces with intellectual life, and in stultifying life into a material force."

A microphone is not an ear, a camera is not an eye and a computer is not a brain. We should not allow ourselves to be so confused or wrapped up in the technology that we fail to assert the importance of human beings.

We have to decide whether we will fight for our right to be the architects of the future, or allow a tiny minority to reduce us to bee-like responses.

LUCAS PLAN SIGNIFICANT

The Corporate Plan is significant in that it is a very concrete proposal put forward by a group of well organized industrial workers who have demonstrated in the past by the products they have designed and built, that they are no daydreamers. It has demonstrated so clearly, to a whole host of scientific and technical workers through the medium of their own jobs, what the limits of the system are. Many of them actually used to believe that the only reason society didn't have nice socially useful products was that nobody had thought of them! The fact these these products are being built and are still being rejected, both by the government and the Company, demonstrates in very dramatic terms the kind of priorities dominant in this society.

SOCIAL INNOVATION

We believe it is arrogant for aerospace technologists to think that they should be defining what communities should have. The Lucas workers are deeply conscious that if the debate were limited to industrial workers of this kind, it would represent a new form of elitism. They are therefore making strenuous efforts to involve wide sections of the community at large in discussions around these issues so that they can interact with them and learn from them. We are seeking, through the local trade unions, political parties and other organizations in each area, to get people to define what they need, and to begin to create a climate of public opinion where we can force the government and the company to act.

To this end, Lucas workers are cooperating with the North East London Polytechnic to convert a coach into one of their hybrid road/rail vehicles.[10] They hope that this vehicle will be a form of technological agitprop. It is intended to have an exhibition within the coach with photographs, slides and videotapes describing the concepts underlying the Corporate Plan and showing some of the prototypes of the products in action. It is hoped that local trade union branches and trades councils, together with community groups, will sponsor the visits of the vehicle to the different cities and that these will culminate in large public meetings where discussions between members of the public and technological and industrial workers can take place.

Part of the exhibition in the vehicle will be a display of photographs composed by Dennis Marshall, a skilled worker at Lucas Aerospace. It vividly demonstrates the way that the ideas embodied in the Corporate Plan have released not only the technical creativity, but also the artistic creativity among the employees. Dennis Marshall has produced beautiful and vivid depictions with his camera, of pollution, decay of inner cities, neglect of railway systems and nuclear hazards. When I used these as illustrations to one of my talks at the Royal College of Art, the people there were amazed that an industrial worker could have produced such impressive work. They were even more amazed when I suggested that we would all benefit if they came to work at Lucas and made way for Dennis Marshall at the R.C.A.!

TRADE UNION RESPONSE

At national level, the trade union movement has given very little support and encouragement to the Corporate Plan, although there have been some positive developments. The Trade Union Congress* (T.U.C.) has, for example, produced a half hour television program dealing with the Plan, and this has appeared on BBC Channell 2 as part of its trade union training program for shop stewards.

The Transport and General Workers' Union has come out with a statement indicating that its shop stewards throughout the country should press for corporate plans of this kind.

At an international level the interest has been truly enormous. In Sweden, for example, they have produced six half hour radio programs dealing exclusively with the Corporate Plan and have made many cassettes which are now being discussed in factories throughout Sweden. They have also made a one hour television program, and a paperback book has been produced in Swedish dealing with the Corporate Plan. Similar developments are taking place in Australia and elsewhere. The interest centers not merely on the fact that a group of workers for the first time are demanding the right to work on socially useful products, but that they are proposing a whole series of new methods of production where workers by hand and brain can really contribute to the design and development of products, and where they can work in a non-alienated manner in a labor process which enhances human beings rather than diminishes them.

---

*The Trade Union Congress is the umbrella organization in Britain that represents some 11.5 million trade unionists.

In the past, our society has been very good at technical invention but very slow at social innovation. We have made incredible strides technologically, but our social organizations are virtually those which existed several hundred years ago. One of the Swedish Television interviewers said, "When one looks at Britain in the past, it has been great at scientific and technological invention and frequently has not really developed or exploited that. The Lucas Workers Corporate Plan shows a great social invention, but it probably is also the case that they will not develop or extend that in Britain. If this were true, it would be very sad indeed."

THE TECHNOLOGICAL ASPECT OF THE PLAN

Although the social and political implications of the Lucas Aerospace Workers Campaign have received considerable attention, the technology contained within the Corporate Plan has largely been ignored, even though the workers themselves in their Plan placed considerable emphasis upon the forms taken by the technology, the products and the manner of producing them. This is particularly true of criticism of the Plan.[11]

This reluctance to deal with the technology of the Plan is, on the one hand, due to the remarkable incompetence of those on the left in the field of science and technology, and on the other, to an indifference to technology because it is perceived, as described earlier, to be "neutral."

The Lucas workers had sought to find, and debated in considerable detail, those forms of technology which would give full vent to the creativity of the hands and minds of the workers, and which could be carried out through non-hierarchal forms of industrial organization. For the fact is that for those at the point of production, the consideration of the technology, the design methodologies, and the nature of the labor process which arises from them, is of equal importance to the political considerations precisely because these workers do not separate one from the other. Indeed, one of the most positive features the Lucas workers see arising from their Corporate Plan are the discussions now taking place with shop stewards and representatives of workers at all levels in industry, from scientists to semi-skilled workers, in a range of companies from Vickers, Parsons, Rolls Royce, Chryslers, Dunlops and Thorns. These discussions center not just around the "political aspects," but are giving rise to a profound questioning of the nature of the technology itself and the design methodologies used.

In the course of actually designing and building prototypes they have discovered, as one worker expressed it, "...that management is not a skill or a craft or a profession but a command relationship; a sort of bad habit inherited from the army and the church."

They have also shown, if only in embryo, that the design methodology used in a "socialist technology" would have to be radically different from that which applies in our current technology.

At present, in the technologically advanced nations, highly qualified designers and technologists spend months drawing, stressing and analyzing a prototype before telling the workers on the shop floor what should be done. These design stages involve rarified, complex mathematical procedures which are necessary only because, for commercial reasons, materials have to be exploited to the full. Both the materials and the systems of the products are designed just to perform a precisely defined function for a very short length of time before the product is rendered redundant (planned obsolescence). The rarified mathematical procedures are outside the experience of workers and are used as a means of bludgeoning their common sense into silence.

DRAMATIC EXAMPLE

There is a tendency for computer specialists to imply that they have the solutions to all our problems without necessarily having much real design experience behind them. Dramatic examples of what can result from this are already coming from the United States. At one aircraft company they engaged a team of four mathematicians, all of PhD level, to attempt to define in a program a method of drawing the afterburner of a large jet engine. This was an extremely complex shape, which they attempted to define by using Coons' Patch Surface Definitions.

They spent some two years dealing with this problem and could not find a satisfactory solution. When, however, they went to the experimental workshop of the aircraft factory, they found that a skilled sheet metal worker, together with a draftsman, had actually succeeded in drawing and making one of these. One of the mathematicians observed, "They may have succeeded in making it but they didn't understand how they did it." This seems to me to be a rather remarkable concept of reality. It does, however, dramatically illustrate the manner in which the three dimensional skill of draftsmen and skilled workers can be thoughtlessly eliminated in this drive to replace people by equipment. All their knowledge of the physical world about them acquired through years of making things and seeing them break and rupture is regarded as insignificant, irrelevant or even dangerous!

With the prototypes developed from the products proposed in the Corporate Plan, the methodology of production was quite the reverse of the above. Workers on the shop floor had every opportunity of giving full vent to their skills and creativity since the prototypes were designed more by "feel" than by "analysis."

There is nothing mysterious about this "feel" or tacit knowledge as I will call it. It comes about as a result of years of direct experience at the point of production itself. It is a sad reflection of the specific form technology takes in this society that this wealth of knowledge is deliberately eliminated. Indeed, this elimination is the very root of so-called "Scientific Management" or Taylorism. Taylor said of his ideas: "...that the worker is told exactly what he is to do and how he is to do it, and any improvement he makes upon the instructions given to him is fatal to success."

Clearly, any talk of industrial democracy with this kind of technology is simply a deception.

AN ELEMENT IN DESIGN

It will undoubtedly be argued by the authoritarians, both of the right and of the left, that the Lucas Workers' approach to technology is romantic, unrigorous and unscientific. Such a view ignores the fact that a desire to meet real social needs is a vitally important stimulus to good quality and creative design, and is a qualitative element of design which cannot be treated in a mathematical way as can the quantitative elements.

Nor are the Lucas workers alone in taking this viewpoint of science and technology. In a recent paper, one of the country's leading technologists and academics, Howard Rosenbrock, had this to say:

> My own conclusion is that engineering is an art rather than a science and by saying this I imply a higher, not a lower status. Scientific Knowledge and mathematical analysis enter into engineering in an indispensible way and their role will continually increase. But engineering also contains elements of experience and judgement and regards all social considerations and the most effective ways of using human labour. These partly embody knowledge which has not yet been reduced to exact and mathematical form. They also embody value judgements which are not amenable to the scientific method.[12]

That this approach to technology is not only more democratic, but also much more fruitful, is demonstrated by the impressive range of products which the Lucas workers have proposed and in some instances are building and testing in prototype. Their approach to the problem of finding an ecologically desirable power unit for cars, coaches and trains illustrates the point quite dramatically.

FIVE

Some Social and Technological Projections

There is now widespread interest in what is vaguely called "community activity". A motley collection of politicians of the Right and Left, church leaders, local councillors, social workers and the inevitable professional planners are all pronouncing in a grandiose pluralism, the need to "revitalize our communities."

In these pronouncements, "community" seems to be taken to mean all those who reside in a given geographical area. Seldom does it seem to be admitted that within "a community" there will be conflicting class and cultural interests. Thus, although it is never explicitly stated, it seems to be assumed that the interests of a working class old age pensioner dying of hypothermia would coincide with those for example, of the owner of a small building firm (small can mean employing under 200 people). Indeed the establishment of small businesses is widely projected as the essential element in community rejuvenation. How and why these small businesses would differ from those in the past is studiously avoided. It seems to be forgotten that some of the most grotesque and hazardous conditions were always to be found in the sweatshops of our inner cities. Rag Trade Workshops* in the East End of London even to this day are vivid reminders that small business does not provide the kind of working conditions to which human beings should aspire in the last quarter of the twentieth century.

A choice between this type of industrial development or the present social disintegration is, of course, an entirely false one. What we should be considering are communally owned enterprises on non-hierarchal lines, and closely integrated to local needs for equipment and services. A much more serious aspect of this "community rejuvenation" is the drift towards what I shall call industrial feudalism. Our economy is now dominated by the massive multinational corporations and financial institutions. The role of even nation-states is quite subordinate to these, since they set the economic, and increasingly the political, framework within which the governments of the individual nation-states are allowed to legislate. These vast corporations spearhead the so-called "technological revolution," and distort its development to meet their needs for profit maximization. The underlying assumption is that of a rapacious economic system fired by ever increasing consumption and production. Apart from the waste of energy and materials their throwaway products bring in their wake (not to mention widespread pollution), they are becoming increasingly capital intensive rather than labor intensive. Throughout the so-called "technologically advanced nations," they are displacing millions of workers and are permanently destroying jobs and skills while extending their control over the cultural as well as the economic and social lives of the mass of the people.

In addition to this, the ecological crisis is likely to assume even greater political significance and it is important that it should not be seen as a middle class preserve. After all, it is always the stream in which a worker does his bit of rough fishing that will be polluted, seldom if ever will it be the salmon

---

*The Rag Trade Workshops are the sweatshops of the clothing industry, where traditionally very low wages and bad working conditions apply.

stream in Scotland. It is always the working class community that will have a freeway running through it, not the stockbroker belt in the suburbs. It is usually the working class playground that will have filth belched out upon it from local factories, hardly ever the middle class golf course. Workers do not stand to gain from the misuse of science and technology. They make no profit from the pollution of the rivers, the seas and the air. It is in their class interests to resist these things and it is vital that they should be involved.

Public hostility to the naked power of this "democratic oppression" is growing and ranges from the reluctance of young people to work for the big multinationals to the more dramatic actions against businessmen in West Germany and Italy. These are as yet isolated indicators of what may become a very significant movement, when large sections of the labor and trade union movement finally realize that in employment terms we are dealing with structural unemployment rather than the cyclical forms we experienced in the past.

The vast multinationals are increasingly conscious that there is going to be a backlash from society at large due to the way they are distorting its development. Some of my acquaintances who are well placed technocrats in these vast companies tell me that they are about to engage in what they call "programs of enlightened self interest," and are planning to move into the community and job creation field.

The idea is that the big firms will supply some funds, and send their executives (with all the elitism that implies) to set up small scale community enterprises. In this way they hope to be able to placate the public on the one hand, but on the other to thereby be left free to get on with the serious business of maximizing their profits at the commanding heights of the economy.

Computerization and automation will mean that smaller numbers will be required to run the large corporations. These will be a separate elite from the rest of the community and highly organized on the "business union" basis as in the United States. They will be true "corporation men," satisfied economically with company cars, company houses and company medicare schemes. There will be funds for the schooling of the corporation man's children, special superannuation schemes and, of course, expense accounts for overseas travel, entertaining and other corporation "responsibilities".

A large sector of those who remain—the unemployed—will be left to fiddle around with "community work". These activities will be deliberately chosen because they yield no economic power. Indeed it will be a sort of theraputic, do-it-yourself social service. The industrial feudal lords will sit in the multinational headquarters while the peasants scratch out a living in the deprived communities in which they reside. In practice this will mean that they will spend their time repairing, cleaning up, modifying and recycling the rubbish which the large corporations are imposing on them. While it is conceivable that some of these jobs will be craft based and thus provide an outlet for some initiative and self activism, the significant reality will be that they have no economic power and no industrial muscle. So a significant proportion of the population seems destined to have no "real job" with all the social multiplier effects which this will give rise to and many of which are already apparent.

We will all be assured, of course, that this kind of work will be non-alienating and will enhance our self reliance. Both of these are highly desirable in themselves, but that will not be the objective of the sort of community activity now envisaged and being supported by the government. The implications of such a development are far more profound than it would first appear.

There is growing evidence that in sectors of the government as well as in the boardrooms of the massive multinational corporations, there is the intention of actively encouraging the growth of this dual economy. Some of the corporations have already released leading executives to become involved in job creation schemes. As one of them jokingly put it, "The Government can only pay about £8,000 for them to do this work so we'll make up the other £30,000 for them!"

Their concern seems to be to diffuse the growing resistance among the unemployed and the critical sectors of the community to the manner in which the corporations and the government are displacing large groups of workers on the one hand and dominating the manner in which technology is developing on the other.

In the process they are, of course, maximizing their profits and in practice are able to circumvent or negate any enlightened programs of industrial reorganization which nation-state governments might attempt to implement. Such enlightenment, I must add, is not particularly evident anywhere in this country under this government or indeed the last one. It is not, therefore, too far fetched to suggest that the government and the large employers will conspire to force the unemployed and the deprived communities to provide their own social services.

Culturally, the members of these communities may live in a world they no longer understand or can cope with. At a political level, with such a concentration of economic power and technical knowhow within the elite, it is unlikely that the present concepts of equality and democracy would long survive, and the development of highly centralized, authoritarian corporate state would thereby be facilitated. Further, the small elite in the highly capitalized "scientific and productive" sector of the economy would be involved in the design and development of forms of repressive technology which could be used against the remainder without the hindrance of any countervailing force in the traditional form of a class conscious, organized working class within the productive processes.

A FOOT IN BOTH CAMPS

It is against this scenario that the Lucas workers have raised the demand that everybody should have the right to a job, and to socially useful work. If there is to be a dual economy with high level and low level sectors, then the work in the advanced sector should be available to and shared out among the whole labor force. No one who wants to take part should be written off as incompetent or incapable, and an appropriate educational and training infrastructure should be made available to all. Socially useful production should raise the level of interest, involvement and job satisfaction of workers and help to release the immense creativity of the workforce which is at present deliberately stifled through Taylorism and Scientific Management. Work sharing would entail a dramatic reduction of work hours in the "productive sector" leaving workers time to engage in the alternative "uneconomic service sector".

This is not as far fetched as it may sound. Probably the majority of the population is already engaged in some form of do-it-yourself or voluntary activity not carried on for profit, whether it be community action, hospital visiting, house repairing or home brewing, and there is great scope for cooperative organization in any such activities where group working is desirable.

The Lucas workers are keen to establish links with workers' cooperatives with whom they could exchange useful products not necessarily suited to large scale production. However, they strongly resist suggestions that their own factory should be turned over to independent cooperative ownership at this present historical stage. This would merely place the present problems of industry directly on the backs of the workers without altering the structure of society or the economies from which these problems arise in the first instance.

All of this implies a significant shift in the way in which trade unions would function. Firstly, it would require them to attach far greater importance to workplace organization than is now the case. The present grudging acceptance of developments such as those at Lucas Aerospace would have to be replaced by active support. Secondly, if we are to have a dual economy, the unions must be prepared to function in both sectors, to organize the unemployed and the partially employed and to expand the cooperative sector where traditionally 100% trade unionism has been readily accepted. To succeed in doing this, a dramatic re-examination of trade union structures will be necessary, and the highly centralized authoritarian bureaucratic forms of some of them would have to be altered. More importantly, they could provide a bridge both through education and collective bargaining whereby workers can as a right have a foot in both camps if we are going to accept that there will be two. In other words, they will be helping to link together the dual role of human beings, namely that of producers and consumers.

ON WORKERS' CONTROL

If we look about us in Britain or the United States at the moment, we can see there is an incredible crisis. There is a crisis of structural unemployment, our social services are being dismantled, and even the air we breathe is being polluted by this system. Yet, in spite of this crisis, the influence of the left is significantly small in both Britain and the U.S.

When you say that to left wing people, they have a tendency in all seriousness to take the view of Brecht who, on one famous occasion said with brilliant irony "The Government has decided the people are wrong, therefore the people must be disbanded." Some people say if *only* we had a French working class, or if *only* we had the Italian working class we would surge forward. But we have the British working class with all its weaknesses and all its strengths—and its strengths are many. It is an experienced and courageous working class.

Part of the trouble is that we don't listen to the working class nearly enough. When you talk to workers about a socialist society, they tend to ask whether there is any country where that sort of society exists, and that's a pretty difficult question. I wouldn't like to say that the sort of society I want is the one I see in the Soviet Union, for example.

They ask "What kind of leaders? How would the country be run?" It is clear from the whole way they question these things, that they have no intention of replacing one elite with another. They don't want tsars, whether they be government tsars, trade union tsars or any other kind. They want a society in which they can really participate and use their creativity to the full.

Now in many ways, that concept is in contradiction with the notion of leadership that exists in many parties and factions in the United Kingdom. There is the notion going around that leadership is declaring yourself to be a vanguard. Having said that, you then pursue a sort of Jesuitical logic and say that if we are the vanguard it necessarily follows that we represent the highest level of consciousness of the working class. We are also the most dynamic in the working class and anybody who disagrees with us must be an enemy. This dogma induces political leaders to perceive their role as solely that of telling others what to do, since by virtue of being part of this elite vanguard, they know by definiton precisely how the working class should behave in all circumstances and at all times.

Now the problem about this is that even if we could find leadership—and I certainly don't see it in Britain at the moment—such a leadership would deny the working class a most precious experience. This is the self-activism and development which raises the level of consciousness and competence to that high level which is a prerequisite of social advance, that is, a really democratic society. Insofar as workers control is one of the components which would do that, in my view, it is important.

WORKERS HAVE VISION

I mention this particularly because of the significant developments in Lucas Aerospace. As a result of trying to draw an image of what the future of that company might be like, and an image of the society in which it would operate, it has been possible to be practically engaged in a whole series of activities which have exposed our members, far more clearly than ever before, to the objective role of the government and many of the ministers within it. We have seen both the pathetic, slavish grovelling of some government officials to this big multinational company in which we work and the objective role of the trade union leaders who, behind our backs, have had secret meetings with the company to prepare the carve up of our jobs. These are the same people who, when confronted two years ago with our request to assist us in deciding what products we could be making and how we should be making them, were absolutely silent. Yet they are the people who will tell us that they know best and will always lead us.

In the course of this development, the whole question of hierarchies and managements has been at the forefront of our discussion. "We have discovered" said one worker, "that management is not a skill, a craft or a profession but a command relationship, a bad habit inherited from the army and the church."

We don't like those kind of hierarchies at all whether they be in the trade union movement or in political parties.

In the course of that struggle, we have as well been able to demonstrate in practice the non-neutrality of science and technology. We have been able to do it in objective circumstances through activity, in a way we could never have done just by reading about it, or by getting lectures from very profound leaders. "I know because I do" as one of the Lucas workers said "not, I do because I know."

Any organization which provides a framework in which workers can be involved in that kind of activity is, in my view, an important and significant development towards creating the level of consciousness we need to safeguard and guarantee democracy in the future.

STAY AT THE BASE

When the superstructure has been changed in other countries, there has been a tendency to run industry precisely as before. It was Lenin who said how important Taylorism could be to the running of the Soviet Union. Now you may recall that Taylor said "In my system the workman is told precisely what he is to do and how he is to do it, and any improvement he makes upon the instructions given to him is fatal to success." It reminds me of some political leaders who claim that the mass has been subverted when it does not pursue the direct party line!

What we are talking about is a level of consciousness which comes through struggle. We maintain that those who become separated from the base of the struggle and go into the superstructure very quickly begin to challenge the base. There is a contradiction between superstructure and base. In my view, any trade union leader who becomes full time—he can be as benevolent and dynamic and energetic and political as he wants—taken away from the point of production, that person, over five years or so, will change. I have seen that change happen among my colleagues. If we are talking therefore about workers control, we have got to ensure that we develop mechanisms where people are continuously exposed to the contradiction at the point of production itself.

When Jack Mundy suggested that trade union leaders should return to the point of production after three years and work there for six years before they have the possibility of becoming a full time official again, he was torn apart both by the right and left in Australia. Both of these groups saw real workers control, where there was rotation of function and involvement of people at the point where the decisions would be made, as a challenge to their power structures. It is worth recalling, that during the Cultural Revolution in China (which I find very difficult to make a final judgement about), the workers at the Shanghai machine tool factory said on one occasion that in their opinion, the most dynamic members of a Party or class should never go into the superstructure, but should stay at the base, fermenting and toppling the super structure if necessary. That to me is the dimension of real industrial democracy and control.

COMPROMISE?

Now it may be that industrial democracy and workers control as it is spoken about, could represent a compromise with the system which oppresses us. There seems to be in some circles the idea that if only we could get more and more people into positions of authority, we might wake up one morning to find that we have five seats on the Board against the employers' four seats and we could then disband it altogether.

Now I don't believe for one moment that any ruling class acquiesces in its own destruction. I take the view that there is a need ultimately for a party and for an organization of the working class which can face up to that power. Although that would require a level of consciousness among the working class which we don't have in Britain at the moment, our experience in Lucas Aerospace is that we are developing the levels of consciousness which will make that sort of thing possible.

I think, therefore, that insofar as workers control can begin to move towards a dual power situation in industry, where the workers begin to flex their own muscles and be conscious of their own great intellects, then it is of some importance. These are the people who design and build

Social and Technological Projections 117

everything we see about us and without whom we could not survive. After all, you can't live in a dollar bill, you can't eat a dollar bill, neither can you drive around in one. All that we see about us comes as a result of the power, the ingenuity and the creativity of working people. If through workers control they have the opportunity of sensing that power, of using it in practice and thereby uderstanding how parasitic and how irrelevant are those who control society, then workers control will have been important. Insofar as it represents one challenge to the naked power of the multinationals in this country, it is, I believe, of great significance.

NUCLEAR POWER—THE POLITICAL DANGERS

Another issue of great significance finally being questioned by the trade union movement today is the acceptability of nuclear power. And although the importance of this beginning should not be underestimated, I regret to have to say, right from the start, that the views I express below on the question of nuclear power are still minority views among my fellow trade unionists. All ideas, however, and all great campaigns start off with minorities. Truth can also lie with minorities, so although I admit that my views on this issue are not shared by most trade union leaders, and the T.U.C. in particular, I don't think that this should worry us. The T.U.C. was wrong when it said the Social Contract would improve our living standards. It, in fact, drastically cut them. They were wrong when they said that government policies would reduce unemployment. They increased it. So one thing we can say about their policies on nuclear power is that at least they are being consistent. For they have got that wrong as well.

So on this issue, as on all the others, it will be up to the rank and file to put them right!

It is not my intention to attempt to deal with the issues of nuclear hazards or the wider ecological issues. I will confine myself to what I regard as the enormous political consequences of this type of technology. These political issues can, and should represent a major rallying point for the trade union movement. Nuclear technology will be a Trojan Horse with which to attack basic democratic and trade union rights, which our movement has established after generations of struggle and sacrifice.

For 300 years, our predecessors fought against the employers, governments and the law to establish the right to strike. Only a slave cannot strike!

This technology will prove to be one of the most effective strike breakers in history. When the former Conservative Government sought to deny us some of these rights through its industrial relations policies, there was a massive upsurge against it, which finally resulted in the defeat of the Heath Government. With this type of technology the same sort of thing will be done to us by much more surreptitious means.

Tony Benn, the Minister of Energy, is an advocate of industrial democracy, yet he threatened to use the troops to end a dispute in the nuclear industry on the grounds that the strike constituted a major public hazard.

Once it has been established through the nuclear industry that it is industrial policy to prevent strikes in circumstances of this kind, you could then extend the argument to many I.C.I. plants, or indeed to almost any large scale plant, as we know from the recent events in the North of Italy. This kind of industry could do to us what anti-trade union legislation has failed to do to us in the past. That is, deny us a most basic right—the right to strike.

BRITISH BERUFSVERBOT

Trade unions have always tried to prevent employers victimizing workers because of their political views. With this type of nuclear technology, it will be said that to guard against terrorists, the government must increasingly insist upon knowing the political affiliations, the intimate personal habits and even the bank balances of our members who work in those industries. It will be a massive intrusion into the personal privacies of our members who are expected to do this work, and workers will be excluded or denied the right to work in such industries on the basis of the political views that they hold. We will no longer be able to sneer at West Germany, with its repressive legislation. We will have our own, much more subtle English form of Berufsverbot.

Social and Technological Projections

Even the communities which live around the stations will be subjected to scrutiny, in case they are such as to harbor potential terrorists.

For those who work in the industry itself, there can be no industrial democracy. Since the industry and its operation represent enormous industrial hazards all actions of the workers in it are predetermined well in advance. All command systems emanate centrally and must be obeyed at all levels. It is run on almost military type lines. Even the clothing that you wear is specified in many of the areas.

One of the latest arguments is to say that we must have this technology because it will create new jobs for us. We suddenly find lots of strange allies of the working class; those converted overnight to concern about our right to work. We heard little from these people when Arnold Weinstock was destroying 60,000 jobs in G.E.C. We heard little from them when companies like Leyland were destroying thousands of jobs as well. In particular, we in Lucas Aerospace heard nothing from them when our company brutally reduced its workforce from 18,000 to 12,000. If for a moment we accept that these people are genuinely concerned about jobs for the working class, we would have to say immediately that their proposal that we need nuclear power to do this, will represent the most expensive job creation scheme in history. Probably in more senses than one.

If we regard this expense, not in terms of hazards or potential loss of life, but simply in first order economic terms, it will cost something like £600,000 to create one permanent job at Windscale. Yet one could create jobs in energy conservation, say in the East End of London, in insulating houses at approximately £4,000 per job.

We should, in fact, be considering very carefully how we use the energy we've got, and how much energy we really need to generate. We should seriously think about using wisely that energy which we already have available. We could conserve enormous amounts of energy, and this is no pipe dream. My colleagues at Lucas Aerospace have proposed a range of socially useful products—over 150 proposals in total, but within the concept of socially useful, we include the conservation of energy and materials and the provisions of jobs which are non-alienating and non-fragmented.

LEARN TO USE THE ENERGY WE HAVE!

One of the proposals is for a unique hybrid power pack which could be used in cars, buses and trains, and could work relatively maintenance free for about 20 years. We reckon that during the course of its operation, it would reduce energy consumption for its specific application by about 50%. Our colleagues in Burnley have designed a natural gas fired heat pump which, when you have zero degrees outside a building, will give you 2.8 times the energy input from the gas in the form of heat inside the house. This is because the pump extracts heat from the atmosphere outside. When its 10°C outside you get four times the energy input.

I wish to state, as a trade unionist and technologist, that I am not opposed to technological change. I am certainly not like some romantics who seem to believe that before the industrial revolution the populace spent its time dancing around maypoles in unspoiled meadows. I am deeply conscious of the enormous contribution science and technology has made toward eliminating squalor, disease and filth. What I am totally opposed to is the irresponsible use of technology, and I regard it as irresponsible to introduce a form of technology such as nuclear power, until we have examined very carefully what real alternatives exist. Until we have put as much research money into alternative forms, as we presently put into nuclear power. Further, that we test and assess the long term implications of some of the nuclear technologies at present being proposed.

Nuclear power as we now know it, will not create the type of jobs we should be demanding in the trade union movement. They will be hazardous jobs. Hazardous for the workers and for the community. There will be enormous political implications in the social infrastructure which will be set up around these industries. It *will* destroy our right to strike.

Those of us who work at the rank and file level in the trade union movement have an enormous task to get these issues raised in that movement, and gradually to get opposition built up to it. There are now many of us attempting to do this, and we join with anti-nuclear activists all over the world in saying— Nuclear Power, no thank you!

ORDINARY PEOPLE

I am frequently asked if I believe that ordinary people are really able to cope with the complexities of advanced technology and modern industrial society.

Firstly, I have never met an ordinary person. All the people I meet are extraordinary. They are fitters, turners, housewives, nurses, airline pilots, doctors, draftspeople, designers, teachers. They all bring to bear vast bands of intelligence, experience and knowledge to the daily tasks that they perform. These "ordinary" people in the performance of their tasks are quite extraordinary!

We have ordinary maintenance fitters who go to London Airport if a generator system is causing a problem. The whole aircraft might perhaps be grounded because of it. One of these fitters can listen to the generator, make a series of apparently simple tests—some of the older fitters will touch it in the way a doctor will touch a patient—and if it is running, will be able to tell you from the vibrations whether a bearing is worn and which one.

The fitter will subsequently make decisions about the reliability of that piece of equipment, and upon that decision, the lives of 400 people may directly depend. The decision may be far more profound in many ways than that which a medical practitioner might make, yet if you asked those "ordinary people" to describe how they reached that decision, they could not do so in the usually accepted academic sense. That is to say, they would not be capable of drawing a decision making tree leading to their final conclusion, yet that conclusion will be right, because they will have spent a lifetime accumulating the skill and knowledge and ability which helps them to arrive at it.

When a great politician goes on a world tour, he or she will be dependent among other things, on the skill and ability of the people who have maintained the aircraft, that of those who designed and built it, the pilots who fly it and the traffic controllers who regulate its flight paths. However, all these will be completely unheard of ordinary people.

Likewise, when we travel on a high speed train, we are dependent on the skill and ability of those who have maintained and built it, the people who maintain the tracks and the people who operate the signalling systems.

In everything we do, whether we are in hospital, travelling along an expressway or in the subway, we depend on the skill, ability, understanding and intelligence of so-called ordinary people. Every skyscraper we see about us has been constructed by people like that. Every car that runs along the roads has been built by them. Yet, in spite of all the knowledge these people demonstrate in practice, they are effectively excluded from the major decisions which affect the way their lives will be run and the industries in which they work organized. They are induced into believing that they are incapable of making the major decisions about the way society should develop.

This in spite of the fact that everything we see about us has been designed and built by those people. They are deliberately conditioned to feel no association between the technology or the products they have produced and their own intelligence.

INTELLIGENCE OR LINGUISTIC ABILITY?

When building workers erect a building they do not scratch their name on it as an artist or a sculptor might do. They do not associate with the building itself, yet it has been produced by them. The whole educational and political system works to reinforce these assumptions. We have more regard for those who write and talk about things than those who actually do them. We confuse linguistic ability with intelligence.

Workers express their intelligence by the things they make and do, and the manner in which they organize themselves. When for example, they are erecting a power station and are communicating with each other on the stages and processes they must go through in doing that, they convey in a few direct, simple words, whole bands of knowledge and experience.

In lining up a turbine and generator set, the workers will go through highly complex decision making routines, and communicate these decisions to each other with crisp simple sentences. If, on the other hand, one were to attempt to write an instruction manual as to how this would be done, it could be a vast work of technical instruction. If you hear two "intellectual workers" talking about these procedures, they frequently have to go into great detail and describe these matters in a language structure of great complexity.

A maintenance fitter at Lucas Aerospace assures me that they have to describe these things at such length and in such a complex fashion that it simply illustrates that they don't quite understand what they are talking about, and therefore have to try and make everything explicit so that they can communicate the ideas to somebody else who likewise lacks the "feel" for what's going on.

It seems therefore to me, that if we are to effectively question the way science and technology are developing, and do it in such a manner that we involve masses of people in the process we shall have to so organize our affairs as to release the tacit knowledge of these workers.

Further, we shall have to organize our decision making routines and our social organizations in such a way that the knowledge, intelligence and experience of these people is not bludgeoned into silence by academic rambling and technological jargon and a deliberately confusing over-complexity.

This is not to say that profound questions can be treated in a simplistic fashion. Rather, it is to say that they should be dealt with in a manner which makes them accessible to so-called ordinary people. These questions are of profound and fundamental importance to the whole issue of democracy itself. If we hold it to be desirable that our societies are composed of alert, vigilant, self active, self reliant, cooperative and concerned citizens, we have to provide political structures which cater to this.

It is my experience, having spent some 25 years in the engineering and manufacturing industries, that "ordinary people" are well capable of understanding and coping with these problems when they are directly put to them. If academics have difficulty in communicating with workers then it is *their* fault, not the fault of the workers involved. As was once said,

> Let your words be so direct and clear and simple that the ideas they represent flow through ordinary people's consciousness as naturally and as easily as the wind and the rain flow through the woods.

COULD WE USE SCIENCE DIFFERENTLY?

I have suggested earlier that science and technology are not neutral, but reflect the economic base which gave rise to them. If this is correct, then the use/abuse model will be inadequate to explain the contradictions we see in technologically advanced society. We shall have to probe deeper.

Technological change has certainly been used, not merely to increase productivity, but to extend control over those who work within those processes. Further, Noble has brilliantly demonstrated that engineers, in the application of science and technology, have been serving and advancing the cause of Corporate Capitalism.[1]

I have questioned whether the means of production so developed would be appropriate in a society where human beings could develop their potential to the full, even when the ownership of the means of production is "in the hands of the people."

Science as practiced in the technologically "advanced" nations, and I would include here the so-called socialist countries, shares with Taylorism the methodological assumptions of predictability, repeatability and quantifiability.

If one accepts these to be tenets of the scientific method, it then follows that to be scientific implies eliminating human judgement, subjectivity and uncertainty; yet skill, in the intellectual and also in the manual sense, can be closely related to the ability to handle uncertainty. Skilled work we may say, is work of risk and uncertainty, whereas unskilled work is work of certainty. The contrast between a skilled turner using a universal lathe and an unskilled worker on a numerical control (N.C.) machine will illustrate the point, as will the contrast between a

Social and Technological Projections 125

conventional designer and one using a "design manual" type of Computer Aided Design System. Further, the exercise of skill is an important learning and developing process. If, therefore, we regard it as desirable to enhance human skill and ability (and I do) then we have to design systems which are responsive to human judgements, and which respond to the persons using them rather than acting upon them. The telechiric devices described earlier begin to address this problem. Other ideas are being explored which in embryo start to point the way to a human enhancing liberatory form of technology. Two examples only will be given here to illustrate the possibilities—one in the field of manual work, the other in the field of intellectual work.

FIRST EXAMPLE

Over the past 200 years, turning has been one of the highly skilled jobs to be found in most engineering workshops. Toolroom turning is one of the most highly skilled jobs of all. The historical tendency, certainly since the war, has been to deskill this function by using N.C. machines. This is done by part programming, a process by which the desired N.C. tool motions are converted to finished tapes. Conventional (symbolic) part programming languages require that a part programmer, upon deciding how a part is to be machined, describes the desired tool motions by a series of symbolic commands. These commands are used to define geometric entities, that is, points, lines and surfaces which may be given symbolic names.

In practice, the part programming languages require the operator to synthesize the desired tool motion from a restircted available vocabulary of symbolic commands. However, all this is doing is attempting to build into the machine the intelligence that would have been exercised by a skilled worker in going through the labor process.

It is possible, by using computerized equipment in a symbiotic way, to link it to the skills of a human being and define the tool motions without symbolic description. such a method is call Analogic Part Programming.[2] In this type of part programming, tool motion information is conveyed in analog form by turning a crank or moving a joystick, or some other hand/eye coordination task, using readout with precision adequate for the machining process. Using a dynamic visual display of the entire working area of the machine tool including the workpiece, the fixturing, the cutting tool and its position, the skilled craftsperson can directly input the desired tool motions to "machine" the workpiece in the display.

Such a system, which may be described as "Part Programming by Doing" would represent a sharp contrast to the main historical tendency towards Symbolic Part Programming. It would require no knowledge of conventional part programming languages, because the necessity to describe symbolically the desired tool motions would be eliminated. This is achieved by providing a system whereby the information regarding a cut is conveyed in a manner closely resembling the conceptual process of the skilled machinist. Thus, it would be necessary to maintain and enhance the skill and ability of a range of craftspeople who would work in parallel with the system.

Significant research has been carried out in this field[3] yet in spite of its obvious advantages, it has not been received with any enthusiasm by large corporations or indeed funding bodies. That this is so would appear to be an entirely "political judgement" rather than a technological one.

SECOND EXAMPLE

In the field of intellectual work, Rosenbrock has questioned the underlying assumptions of the manner in which we are developing computer aided design systems. He charges firstly, that the present techniques fail to exploit the opportunity which interactive computing can offer. The computer and the human mind have quite different but complementary abilities. The computer excels in analysis and numerical computation. The human mind excels in pattern recognition, the assessment of complicated situations and the intuitive leap to new solutions. If these different abilities can be combined, they amount to something much more powerful and effective than anything we have had before.

Social and Technological Projections   127

Rosenbrock objects to the "automated manual" type of system, since it represents as he says "a loss of nerve, a loss of belief in human abilities, and a further unthinking application of the doctrine of the 'Division of Labor'."[4]

As in the case of turning described above, Rosenbrock sees two paths open in respect to design. The first is to accept the skill and knowledge of the designer, and to attempt to give designers improved techniques and improved facilities for exercising their knowledge and skill. Such a system would demand a truly interactive use of computers in a way that allows the very different capabilities of the computer and the human mind to be used to the full.

The alternative to this he suggests is "to subdivide and codify the design process, incorporating the knowledge of the existing designers so that it is reduced to a sequence of simple choices."[5] This, he points out, would lead to a deskilling, such that the job can be done by a person with less training and less experience.

Rosenbrock has demonstrated the first human enhancing alternative by developing a C.A.D. system with graphic output to develop displays from which the designer can assess stability, speed of response, sensitivity to disturbance and other properties of the system.

If, having looked at the displays, the performance of the system is not satisfactory, the displays will suggest how it may be improved. In this respect the displays carry on the long tradition of early pencil and paper methods, but, of course, they bring with them much greater computing power. Thus, as with the lathe and the skilled turner, so also with the V.D.U. and the designer, possibilities exist of a symbiotic relationship between the worker and the equipment. In both cases, tacit knowledge and experience is accepted as valid and is enhanced and developed.

In Rosenbrock's case, it was necessary to examine the underlying mathematical techniques involved in Control Systems Design.[6] The outcome of his work does demonstrate in embryo that there are other alternatives if we are prepared to explore them, and he has suggested that we are now at a unique historical turning point when we may close off options which are now open to us. This process he describes as the "Lushai Hills Effect."[7]

HUMAN ENHANCING

These examples have been quoted to demonstrate that it is possible to so design systems as to enhance human beings rather than to diminish them and subordinate them to the machine. It is my view that systems of this kind, however desirable they may be, will not be developed and widely applied since they challenge power structures in society. Those who have power in society, epitomized by the vast multinational corporations, are concerned with extending their power and gaining control over human beings rather than with liberating them. Such a viewpoint may appear abnormally sectarian or political to some readers, yet they only have to look at the headlines of the technical press to see statements which completely reinforce that analysis. Thus, recently, a headline in the *Engineer* stated "People are trouble, but machines obey"[8] and even in economic papers one finds headlines such as "Robots Don't Strike."[9]

It is not suggested here that engineers who design conventional systems are hideous fascists who deliberately engage in this design in order to subordinate others to the control of the machine and the organizations that own the machine. What I am suggesting, however, is that they are dangerously mistaken in regarding their work as being neutral. Such a naive view was ruthlessly exploited in the Third Reich as Albert Speer pointed out in his book *Inside the Third Reich*:

> Bascially, I exploited the phenomenon of the technician's often blind devotion to his task. Because of what appeared to be the moral neutrality of technology, these people were without any scruples about their activities."

Science and technology are not neutral, and we must at all times expose their underlying assumptions. We can, at the same time, begin to indicate how science and technology might

Social and Technological Projections 129

be applied in the interests of people as a whole, rather than to maximize profits for the few.

The choices are essentially political and ideological rather than technological. As we design technological systems, we are in fact designing sets of social relationships, and as we question those social relationships and attempt to design systems differently, we are then beginning to challenge in a political way power structures in society.

The alternatives are stark. Either we will have a future in which human beings are reduced to a sort of bee-like behavior, reacting to the systems and equipment specified for them; or we will have a future in which masses of people, conscious of their skills and abilities in both a political and technical sense, decide that they are going to be the architects of a new form of technological development which will enhance human creativity and mean more freedom of choice and expression rather than less. The truth is, we shall have to make the profound political decision as to whether we intend to act as architects or behave like bees.

Chapter 1 Identifying the Problem

1. Cooley, M.J.E. "The Knowledge Worker in the 1980s," Doc EC35, Diebold Research Programme, Amsterdam, 1975.
2. Cooley, M.J.E. *Computer Aided Design—Its Nature and Implications*, A.U.E.W.—TASS, 1972.
3. Braverman, Harry. *Labor and Monopoly Capital. The Degradation of Work in the 20th Century*. Monthly Review Press, New York, 1974.
4. Maver, T.W. *Democracy in Design Decision Making C.A.D.* '76. I.P.C. Science and Technology Press, Guildford, Surrey, 1972.
5. Polanyi, M. "Tacit Knowing: its bearing on some problems of Philosophy." Review of Modern Physics, Vol. 34, pp. 601-605, Oct. 1962.
6. Reported in *Daily Mirror*, 7 June 1973.
7. Cooley, M.J.E. Ph.D. Thesis, N.E. London Polytechnic (unpublished).
8. Licklider, J.D.R. "Man-Machine Symbiosis," I.R.E. Trans Electron 2, 1960, p. 4, 11.
9. Bernholz, A. Proc C.A.D. Conference, I.F.I.P., Eindhoven, 1973.
10. Boguslaw, R. *The New Utopians: a study of Systems Design and Social Change*. Englewood Cliffs, N.J., Prentice-Hall, 1965.
11. Taylor, F.W. *On the Art of Cutting Metals*, 3rd Edition Revised. A.S.M.E., New York, 1906.
12. Reported in Dataweek, 29 January 1975.
13. Fairbairn, W. Quoted by J.B. Jefferys, *The Story of Engineers*, Lawrence & Wishart, 1945, p. 9.

Chapter 2 The Human/Machine Interaction

1. Shakel, B. The Ergonomics of the Man/Computer Interface Proc Conf Man/Computer Communication p. 17, Infotech International Ltd., Maidenhead, U.K., November 1978.
2. Faux, R. *The Times*, 26 March 1975.
3. Rose, S. *The Conscious Brain*, Penguin Books, 1976.
4. Archer, L.B. *Computer Design Theory and the Handling of the Qualitative*. Royal College of Art, London, 1973.
5. Nadler, G. "An Investigation of Design Methodology Management," Science, Vol. No. 3, June 1967, pp. 642-655.
6. Lobell, J. "Design and the Powerful Logics of the Mind's Deep Structures," DMG/DRSJ, Vol. 9, No. 2, pp. 122-129.
7. Beveridge, W.I.B. *The Art of Scientific Investigation*, Mercury Books, London, 1961.
8. Eisley, L. *The Mind as Nature*, Harper & Row, New York, 1962.
9. Fabun, D. "You and Creativity," Kaiser Aluminum News, Vol. 25, No. 3.

10. Marx, K. *Capital*, Vol. I, p. 174, Lawrence & Wishart, London, 1974.
11. Silver, R.S. "The Misuse of Science," New Scientist, Vol. 166, 956, 1975, p. 555.
12. Rose, S. "Can Science be Neutral?" Proc Royal Institute, Vol. 45, London, 1973.
13. Rose, H. & S. "The Incorporation of Science," in *The Political Economy of Science*, Eds. Rose & Rose, Macmillan, London, 1976.
14. Shiftworking and Overtime Practices in Computing. Rep Computer Economics Ltd., Richmond, Surrey, 1974.
15. Mott, P.E. *Shiftwork: The Social, Psychological and Physical Consequences*. Ann Arbor, 1975.
16. Rosenbrock, H.H. "The Future of Control," Automatica, Vol. 13, 1977.
17. Bernholz, A. *Op. cit.*
18. Proceedings from Human Choice and Computers. Report HCC, Lp. 5, I.F.I.P., Vienna, 1974.
19. Kling, R. "Towards a People Centred Computer Technology," Proceedings from Assoc. Computer Mach. Nat. Conf., 1973.
20. Ostberg, O. "Review of Visual Strain with special reference to microimage reading," International Micrographics Congress, Stockholm, September, 1976.
21. Report from *New York Times*, Survey NIOSH, New York, 1976.
22. Allen, B. "Health Risks of Working with V.D.U.s," Computer Weekly, 9 February 1968, p. 3.
23. Ostberg, O. "Office Computerisation in Sweden: Worker participation, workplace design considerations and the reduction of visual strain," Proceedings from NATO Advanced Studies Institute on Man/Computer Interaction, Athens, September 1976.
24. Arbeitsbeanspruching und Augenbelastung an Bildschirmgeraten. Rep OGB, Vienna, 1975.
25. Checklist for the Installation and use of Visual Display Units. General and Municipal Workers Union, London, 1978.
26. "Making sure Technology is right for the Press," Computing, p. 74, 23 March 1978.
27. "Electronic Office System designed to Improve Managers' Productivity." Computer Weekly, 21 December 1978, p. 12.
28. Act relating to Worker Protection and Working Environment. Order No. 330, Statens Arbeidstilsyn Direktoratet, OSLO.
29. Urquart, A. *Familiar Words*, cited in Marx, K., *Capital*, Vol. 1, p. 36e, London, 1961.
30. Smith, A. *The Wealth of Nations*, Random House, New York, 1937

31. Martyn, H. *Consideration upon the East India Trade*, London, 1801.
32. Braverman, H. *Op. cit.*
33. Dochery, P. "Automation in the Service Industries," Round Table Discussion, I.F.A.C., 1978.
34. Kraft, P. *Programmers and Managers—the Routinization of Computer Programming in the United States*, Springer Verlag, Berlin, Heidelberg, New York, 1977.
35. Babbage, C. *On the Economy of Machinery and Manufactures*, New York, 1963, (reprint).
36. Carlson, H.C. in Braverman, *op. cit.*
37. Reported in Academy of Management Journal, Vol. 17, No. 2, p. 306.
38. Reported in Management Science, Vol, 19, No. 4, p. 357.
39. Reported in Times Higher Educational Supplement, 14 February 1975, p. 14.
40. Reported in New Scientist, 22 April 1976, p. 178.
41. Reported in *The Guardian*, 12 October 1979.

Chapter 3 Political Implications

1. Marglin, S. "What Do Bosses Do?" in *The Division of Labour*, ed. A. Gorz, Harvester Press, Sussex, 1976.
2. Hoos, I. "When the Computer takes over the Office," harvard Business Review, Vol. 38, No. 4, 1960.
3. Polanyi, M. *Op. cit.*, (Chapt. 1).
4. Sohn-Rethel, A. *Intellectual and Manual Labour—a Critique of epistomology*, Macmillan Press Ltd., London, 1978.
5. Bodington, S. *Science and Social Action*, Allison and Busby Ltd., London, 1979.
6. Bodington, S. *Op. cit.*
7. Needham, J. "History and Human Values," in H. & S. Rose, *The Radicalisation of Science*, Macmillan, London, 1976.
8. Rose, H. & S. "The Incorporation of Science," *Op. cit.*
9. Rose, H. & S. in *The Social Impact of Modern Biology*, W. Fuller, ed. Routledge, Kegan & Paul, London, 1971.
10. Silver, R.S. "The Misuse of Science," *Op. cit.*
11. Henning, D. Bericht 74-09. Berlin Technical University, 20 January 1974.
12. Jungk, R. Qualitat des Lebens, EVA, Cologne, 1973.
13. Braverman, H. *Op. cit.*
14. Lenin, V.I. "The Immediate Tasks of the Soviet Government (1918) in *Collected Works*, Vol. 27, Moscow, 1965.
15. Marx, K. *Critique of the Gotha Programme*, ed. C.P. Dutt, Lawrence & Wishart, London, 1938.

Chapter 4 Drawing up the Corporate Plan at Lucas Aerospace

1. Fletcher, R. Guided Transport Systems, N.E. London Polytechnic, 1978.
2. Reported in The Engineer, 14 September 1978, p. 24-25.
3. Marglin, S. "What Do Bosses Do?", *Op. cit.*
4. Braverman, H. *Op. cit.*
5. Glegg, A. "Craftsmen and the Origin of Science," Science and Society, Vol. XLIII, No. 2, Summer 1979, p. 186-201.
6. Rosenbrock, H. "The Future of Control," Automatica, Vol. 13, 1977.
7. Rosenbrock, H. "Interactive Computing: A New Opportunity," Control Systems Centre Report No. 338, UMIST, September, 1977 and "The Future of Control," *Op. cit.* Rosenbrock's interactive graphic systems are described in his book together with their basic mathematical techniques.
8. Weizenbaum, J. "On the Impact of the Computer on Society—How does one insult a Machine?" Science, Vol. 176, 1972, pp. 609-614, and *Computer Power and Human Reason*, W.H. Freeman & Co., San Francisco, 1976.
9. Cooley, Friberg, Sjoberg. Altervativ Produktion, Liberforlag. Stockholm, 1975.
10. Fletcher, R. *Op. cit.*
11. Albury, D. "Alternative Plans and Revolutionary Strategy," in International Socialism, 6, Autumn 1979.
12. Nadler, G. *Op. cit.*

Chapter 5 Some Social and Technological Projections

1. Noble, D.F. *America by Design*, Alfred A. Knopf. New York, 1977.
2. Gossard, D. and von Turkovich, B. "Analogic Part Programming with Interactive Graphics, Annals of the C.I.R.P., Vol. 27, January 1978.
3. Gossard, D. Analogic Part Programming with Interactive Graphics. Ph. D. Thesis, M.I.T., February 1975.
4. Rosenbrock, H.H. "The Future of Control," *Op. cit.* (Chapt. 2).
5. Rosenbrock, H.H. Interactive Computing: a New Opportunity. Control Systems Centre Report No. 388, U.M.I.S.T., 1977.
6. Rosenbrock, H.H. *Computer Aided Control System Design*, Academic Press, London, New York, San Francisco, 1974.
7. Rosenbrock, H.H. "The Redirection of Technology. I.F.A.C. Symposium: criteria for selecting appropriate technologies under different cultural, technical and social conditions. Bari, Italy. May 1979.
8. The Engineer, 14 September 1978, p. 24-25.
9. The Economist, 14 July 1973, p. 71.